CAMBRIDGE LIBRARY COLLECTION

Books of enduring scholarly value

Astronomy

From ancient times, humans have tried to understand the workings of the world around them. The roots of modern physical science go back to the very earliest mechanical devices such as levers and rollers, the mixing of paints and dyes, and the importance of the heavenly bodies in early religious observance and navigation. The physical sciences as we know them today began to emerge as independent academic subjects during the early modern period, in the work of Newton and other 'natural philosophers', and numerous sub-disciplines developed during the centuries that followed. This part of the Cambridge Library Collection is devoted to landmark publications in this area which will be of interest to historians of science concerned with individual scientists, particular discoveries, and advances in scientific method, or with the establishment and development of scientific institutions around the world.

History of the Royal Astronomical Society, 1820–1920

Founded as the Astronomical Society of London in 1820, this illustrious organisation received its royal charter in 1831. It has counted some of the world's greatest astronomers among its members, most notably its first president, Sir William Herschel, whose family archive forms part of its extensive library. Now based in Burlington House in Piccadilly, it continues to publish journals, award medals and prizes, and support education and outreach work. Following the society's centenary, this survey of its history appeared in 1923 and comprises contributions from leading astronomers of the early twentieth century. The extracts from primary sources include the diary entry of Sir John Herschel, son of William, recording the dinner at which the society's formation was discussed. The work also provides insights into how the society was able to take advantage of imperial expansion to collect observations and data from around the world, fuelling the Victorian pursuit of scientific knowledge.

Cambridge University Press has long been a pioneer in the reissuing of out-of-print titles from its own backlist, producing digital reprints of books that are still sought after by scholars and students but could not be reprinted economically using traditional technology. The Cambridge Library Collection extends this activity to a wider range of books which are still of importance to researchers and professionals, either for the source material they contain, or as landmarks in the history of their academic discipline.

Drawing from the world-renowned collections in the Cambridge University Library and other partner libraries, and guided by the advice of experts in each subject area, Cambridge University Press is using state-of-the-art scanning machines in its own Printing House to capture the content of each book selected for inclusion. The files are processed to give a consistently clear, crisp image, and the books finished to the high quality standard for which the Press is recognised around the world. The latest print-on-demand technology ensures that the books will remain available indefinitely, and that orders for single or multiple copies can quickly be supplied.

The Cambridge Library Collection brings back to life books of enduring scholarly value (including out-of-copyright works originally issued by other publishers) across a wide range of disciplines in the humanities and social sciences and in science and technology.

History of the
Royal Astronomical Society,
1820–1920

EDITED BY J.L.E. DREYER
AND H.H. TURNER

CAMBRIDGE
UNIVERSITY PRESS

CAMBRIDGE
UNIVERSITY PRESS

University Printing House, Cambridge, CB2 8BS, United Kingdom

Published in the United States of America by Cambridge University Press, New York

Cambridge University Press is part of the University of Cambridge.
It furthers the University's mission by disseminating knowledge in the pursuit of
education, learning and research at the highest international levels of excellence.

www.cambridge.org
Information on this title: www.cambridge.org/9781108068604

© in this compilation Cambridge University Press 2014

This edition first published 1923
This digitally printed version 2014

ISBN 978-1-108-06860-4 Paperback

HISTORY

OF THE

ROYAL ASTRONOMICAL SOCIETY
1820–1920

The Rev. WILLIAM PEARSON, LL.D., F.R.S.
(1767–1847)

HISTORY

OF THE

ROYAL ASTRONOMICAL SOCIETY

1820–1920

EDITED BY

J. L. E. DREYER, M.A., Ph.D., D.Sc.

PRESIDENT 1923-

AND

H. H. TURNER, M.A., D.Sc., D.C.L., F.R.S.

PAST PRESIDENT

SAVILIAN PROFESSOR OF ASTRONOMY, OXFORD

WITH CHAPTERS BY THEM AND BY

R. A. SAMPSON, M.A., D.Sc., F.R.S.

PAST PRESIDENT

ASTRONOMER ROYAL FOR SCOTLAND

THE LATE COLONEL E. H. GROVE-HILLS, C.M.G., D.Sc., F.R.S.

PAST PRESIDENT

H. F. NEWALL, M.A., D.Sc., F.R.S.

PAST PRESIDENT

PROFESSOR OF ASTROPHYSICS, CAMBRIDGE

AND

H. P. HOLLIS, B.A.

FORMERLY ASSISTANT AT THE ROYAL OBSERVATORY, GREENWICH

PUBLISHED BY THE

ROYAL ASTRONOMICAL SOCIETY

BURLINGTON HOUSE, LONDON, W. 1

AND SOLD BY

WHELDON & WESLEY, LIMITED

2, 3, & 4 ARTHUR STREET, NEW OXFORD STREET, LONDON, W.C. 2

1923

PREFACE

THE idea of compiling this *History* was started some years before the Centenary. A Committee was appointed to deal with the matter, and it was decided to distribute the work among ten Fellows of the Society, each being responsible for one decade. It was hoped that by beginning thus early the different collaborators would be able to collect materials in a leisurely manner. But the best scheme has its drawbacks, and it is a familiar fact that having plenty of time may result in being late after all. Moreover, in some cases those who had undertaken a share found that their hearts failed them, and it is due to the untiring assistance of Dr. Dreyer, who came to the rescue, that the scheme has been finally carried out in a somewhat modified form. Not only has Dr. Dreyer dealt in all with fifty years out of the hundred, but he has acted as co-editor for the whole, and if I venture to sign my name to this preface as original editor, it is chiefly in order that I may express more fully my grateful thanks to him for all that he has done, which included the important but tiresome undertaking of compiling the index.

The list of independent authors ultimately stands as follows :—

1820–1830,	H. H. Turner .	. pp.	1–49
1830–1840,	J. L. E. Dreyer	. ,,	50–81
1840–1850,	R. A. Sampson	. ,,	82–109
1850–1860,	E. H. Grove-Hills	. ,,	110–128
1860–1870,	H. F. Newall .	. ,,	129–166
1870–1880,	H. P. Hollis .	. ,,	167–211
1880–1920,	J. L. E. Dreyer	. ,,	212–249

It was almost inevitable that in spite of every desire to the contrary some things should be overlooked until too late. One or two points concerning the early years were caught in time to add them on pages 48 and 49; but the later limit also brings its difficulties. The century of which this is the history closed at a time when the Society was again in full vigour and growth after the difficult years of the great war, and some things which occurred after the limiting date were the natural outcome of those which preceded it. Thus it seems proper to mention, even though only in a footnote, the generous donations prompted by the deplorable effects of the war on our finances; and the very sight of this note (on p. 246) suggested (although too late for proper treatment in that place) that the noble bequest of a library of early mathematical and astronomical books, with £250 towards the expenses of housing them, which we owe to the late Colonel

Grove-Hills, should not be ignored because it fell just outside our limiting dates.

Colonel Grove-Hills was treasurer at the time when our centenary was celebrated in 1922, and was actively engaged in rectifying our finances ; his death a few months later was a real tragedy. He is also one of the joint authors of this volume, and it seems in every way appropriate to include his among the few portraits in the volume.

As regards the other portraits, we are fortunate to have recovered one of Dr. Pearson by the kind help, first, of Admiral Sir H. E. Purey Cust, and secondly, of the Rev. R. M. B. Bryant, the present Rector of South Kilworth, who lent us a framed portrait presented to the South Kilworth Reading Room in 1902, by Col. W. Pearson, J.P., D.L., a nephew of our founder. A fine portrait of Francis Baily has long hung in our Council Room, so that his features are familiar to many of us, although those of Dr. Pearson were hitherto unknown ; but a portrait of Baily could not possibly be omitted from this volume. The portrait of John Herschel is well-known as an engraving, but by the kindness of the Master of St. John's College, Cambridge, Dr. R. F. Scott, we are enabled to include a photograph, made directly from the picture by Pickersgill which hangs in St. John's College. The steel engravings of Adams and Airy have been very courteously supplied by Messrs. Macmillan, and will be recognised as having originally appeared in *Nature*. Reference is made on pages 27, 28 to the tragic history of one of our Fellows, Waterston, whose greatness was not recognised in his lifetime ; it seems appropriate to make some poor amends by including him among those whose portraits are given. For De Morgan, we look in vain among the Presidents in the Meeting-Room : we have it on his son's authority that he consistently declined the office on the ground that a President of the R.A.S. should be " either an actual star-gazer or at least a telescope-twiddler." His portrait is here reproduced from the *Memoir* of him published by Messrs. Longmans, Green & Co., with their kind permission, and also that of Messrs. Speaight, Ltd., who now own the original negative taken by Mayall. There is an old print of Sir Joseph Banks, dated 1789, the reproduction of which was debated, because this sturdy opponent of our foundation is shewn holding in his hand Russell's drawing of the moon ; but the ultimate decision was in the negative. The other selections require perhaps no explanation, though apology may be offered for the inevitable defects of any limited choice among so many worthy figures.

H. H. TURNER.

UNIVERSITY OBSERVATORY, OXFORD,
1923 *October.*

CONTENTS

	PAGE
1820–1830, by H. H. TURNER	1
1830–1840, by J. L. E. DREYER	50
1840–1850, by R. A. SAMPSON	82
1850–1860, by E. H. GROVE-HILLS	110
1860–1870, by H. F. NEWALL	129
1870–1880, by H. P. HOLLIS	167
1880–1920, by J. L. E. DREYER	212
INDEX	254

LIST OF ILLUSTRATIONS

Rev. W. PEARSON	*Frontispiece*	
Sir J. HERSCHEL	*to face p.*	8
J. J. WATERSTON	,, ,,	26
Sir J. SOUTH	,, ,,	54
FRANCIS BAILY	,, ,,	90
J. C. ADAMS	,, ,,	96
Rev. R. SHEEPSHANKS	,, ,,	118
A. DE MORGAN	,, ,,	140
Sir W. HUGGINS	,, ,,	152
Adm. W. H. SMYTH	,, ,,	202
Sir G. B. AIRY	,, ,,	210
Colonel E. H. GROVE-HILLS	,, ,,	246

HISTORY OF THE
ROYAL ASTRONOMICAL SOCIETY

CHAPTER I

THE DECADE 1820–1830.   (By H. H. Turner)

WE are indebted to Sir * John Herschel for keeping a diary which begins almost providentially with the year 1820, and from which the Herschel family have kindly made the following extracts :—

Sat., Jan. 1.—Dined with Peacock and Babbage at Provost Goodall's at Eton, & met Col. Thackeray, Vice-Provost Roberts, Capt. Roberts (R.N.), etc.

Sun., Jan. 2.—Peacock and Babbage left Slough after spending a few days here.

Wed., Jan. 12.—Dine (*sic*) at the Freemason's Tavern to meet Dr. Pearson & other gentlemen to consider of forming an *Astronomical* Society.

Sat., Jan. 15.—To attend Committee of yᵉ Astronomical at Geol. Soc. Rooms at 10 o'clock.

Evening.—Mr Lowry's to tea to meet Wollaston, Babbage, Gompertz, Perkins, & Col. Fairman.

Mon., Jan. 17.—Went to Dr. Pearson's, Sheen, near Richmond, where I dined & spent the night.   At night took Mars' diamʳ by way of trying Dr. Pearson's Double image Micrometer.

Tues., Jan. 18.—Spent morning at Dr. Pearson's.   Babbage came about 1ʰ.   Read over & arranged address for circulation with the notice of formation of yᵉ Astronomical Soc.   Dined & returned with Dr. P. and Babbage to the meeting of the Cᵗᵉᵉ in the Evening.

Thurs., Jan. 20.—Returned in the afternoon to Slough.

Mon., Jan. 24.—Death of the Duke of Kent.

Wed., Jan. 26.—Despatched letters concerning the Astronomical Society (the Address & a Circular letter) to Catton, Peacock, & Fallows.

Sat., Jan. 29.—Death of the King.

---

* Knighthood was not conferred till 1831 (followed by Baronetcy in 1838), but for simplicity the familiar name has been used throughout.

A few later extracts are deferred for use in other connections. From the above we gather the important fact that the preliminary meeting was held either before or after dinner at the Freemason's Tavern on January 12. Many years later De Morgan rejoiced to find other evidence of this dinner, which had become almost legendary : the diary shows that it was a pleasant addition to the formal business of the preliminary meeting, fully preserved for us in the Minutes, which may now be given verbatim :—

MINUTES OF THE GENERAL MEETINGS OF THE ASTRONOMICAL
SOCIETY OF LONDON

Wednesday, January 12, 1820

On this day several gentlemen, whose names are hereafter mentioned, met together by appointment, at the Freemason's Tavern, Great Queen Street, Lincoln's Inn Fields, London, to take into consideration the propriety and expediency of establishing a Society for the encouragement and promotion of Astronomy. The following are the names of the gentlemen present :—

No. 1. Charles Babbage, Esq., M.A., F.R.S.L. & E., No. 5 Devonshire Street, Portland Place.
    2. Arthur Baily, Esq., No. 6 Gower Street.
    3. Francis Baily, Esq., F.L.S., Gray's Inn.
    4. Capt. Thomas Colby, of the Royal Engineers, LL.D., F.R.S.E., Tower.
    5. Henry Thomas Colebrooke, Esq., F.R.S. & S.A., Albany, Piccadilly.
    6. Olinthus G. Gregory, LL.D., Woolwich.
    7. Stephen Groombridge, Esq., F.R.S., Blackheath.
    8. John Fred. W. Herschel, Esq., M.A., F.R.S.L. & E., Slough.
    9. Patrick Kelly, LL.D., Finsbury Square.
    10. Daniel Moore, Esq., F.R.S., S.A., and L.S., Lincoln's Inn.
    11. Rev. William Pearson, LL.D., F.R.S., East Sheen.
    12. James South, Esq., 11 Blackman Street, Southwark.
    13. Charles Stokes, Esq., F.S.A. & L.S., Gray's Inn.
    14. Peter Slawinski, D.P. Proff. University, Wilna.

The following mutual agreement was then drawn up and signed by all the gentlemen present, viz. :—

" At a meeting held this twelfth day of January 1820, at the Freemason's Tavern, London, to take into consideration the advantages that are likely to result from the establishment of a Society for the cultivation of Astronomy,—We, whose names are hereunto subscribed, being fully aware of the utility of such an institution, do hereby mutually agree to constitute ourselves a Society, to be called the Astronomical Society of London ; and to be guided, in our future proceedings, by such rules and regulations as may be formed for such Society, in the manner to be appointed at the present meeting for that purpose."

Resolved unanimously—

1. That a Committee of eight members be appointed to draw up such rules and regulations ; and that three be a quorum.

Resolved unanimously—

2. That C. Babbage, Esq. ; F. Baily, Esq. ; Capt. T. Colby ; H. T. Colebrooke, Esq. ; Dr. Gregory ; J. F. W. Herschel, Esq. ; D. Moore, Esq. ; and Rev. Dr. Pearson be the Committee above mentioned.

Resolved unanimously—

3. That a general meeting of the members take place on Tuesday, February the 8th, at the house of the Geological Society in Bedford Street, Covent Garden, at 7 o'clock in the evening precisely ; to take into consideration the rules and regulations which may be then proposed by the Committee.

Resolved unanimously—

4. That any person, recommended by one of the present members of the Society, who may be desirous of joining the Society at, or prior to, the above-mentioned general meeting, shall, on previously signifying in writing his assent to these resolutions, or on authorising a member by letter to signify the same on his behalf, be considered a member thereof without ballot.

Resolved unanimously—

5. That the Committee be authorised to draw up an Address, explanatory of the motives and object of the Society ; and to circulate it in such manner as they may think fit.

Resolved unanimously—

6. That F. Baily, Esq., be Secretary *pro tempore*.

Memorandum.—It was omitted to be stated, in its proper place, that D. Moore, Esq., was unanimously called to the Chair.

FRANCIS BAILY, Secretary *pro tem.*   DAN. MOORE, Chairman.

Returning to the Diary of Sir John Herschel, we see that the first action of the infant Society was the preparation of an Address, and that it was undertaken by Herschel himself, possibly with the help of Pearson and Babbage. The MS. was probably handed to the Secretary, Francis Baily, on the evening of Wednesday, January 19, and on the following Saturday he was able to write announcing that it was in type :—

GRAY'S INN, *Jan.* 22, 1820.

DEAR SIR,—I think you will say that the printer and myself have managed admirably well in being able to decipher and arrange the very rough copy which you left us. There was but one passage which I could not exactly make out ; but as the meaning was evident, I was at no loss to complete the sentence.

There is one liberty I have taken with it, which is the insertion of that passage in page 5 which you had struck out of the MS., but which I thought was too excellent to be thus discarded. On the whole I think the Address admirably adapted to its purpose, and I have no doubt it will be read with pleasure and profit by every lover of Astronomy.

The proof has been finished only this morning, and I have ordered 250 copies to be struck off this afternoon, in order that they may be dried to-morrow ; and, after lying in the press on Monday, be ready for distribution on Tuesday. Had I waited for the return of the enclosed proof, the *whole* of that time would have been lost ; and I think it of material consequence to circulate them as soon as possible. Some will be sent to Scotland, and there will be scarcely time to receive an answer prior to the general meeting, if the printing had been delayed.

However I have thought it better to send you the enclosed copy, in order that you might inspect it prior to its publication : so that, if it did not meet your entire approbation, it might be re-printed, with any alterations you may think advisable. For my own part (and I speak also the opinion of many of the members), I think it cannot be mended.

I shall be obliged by an answer, as early as convenient, addressed to me at the *Stock Exchange* ; and at the same time you will be good enough to inform me, by what conveyance it would be most advisable to forward such copies of the Address and of the Circular as you may wish for distribution.

With my compts. to Sir William and Lady Herschel, believe me, yours truly, Dr Sir,   FRANCIS BAILY.

5 o'clock.—I have waited till the last moment, expecting the printer to send me a copy of the Address ; but, being disappointed, I have despatched this letter without it : to know whether you would prefer seeing it before it is circulated ; or whether I shall forward them as at first proposed ?

On the Monday, Baily was able to send a dozen copies of the Address, but he probably did not get the approval for which he hoped ; for though Sir John's reply has not been preserved, the following sentences from a letter to Babbage dated 1820 February 2, sufficiently indicate his views on the printing of the Address.

Baily made sad work of that scrawl of an address I left with him—it was, to be sure, in great measure my fault, but here and there he has totally inverted the sense of it and made me say what my soul abhors. However, it is the address of the whole Com-mittee, and nonsense distributed among so many will lie lightly on each. However, if you have any copies by you still undis-tributed, do in pity strike out the word " not " in page 8 line 12 in them before you send them abroad.

To identify the offending passage it is necessary to refer to a copy of the octavo pamphlet in which the Address originally appeared. It will serve a double purpose to quote the whole sentence, italicising the word to be struck out.

> One of the first great steps towards an accurate knowledge of the construction of the heavens, is an acquaintance with the individual objects they present ; in other words, the formation of a complete catalogue of stars and other bodies, upon a scale infinitely more extensive than any yet undertaken ; and to be carried down to the minutest objects, *not* visible in any but the very best telescopes.

It is rather puzzling to find that it is impossible to strike out the word *not* as directed if the sense of the passage is to be preserved. The emendation adopted in *Mem.*, **1,** 4, is :

> and that shall comprehend the most minute objects visible in good astronomical telescopes.

The whole construction has, of course, been altered. Sir John's complaint seems to have been a little hasty, for comparison of the Address as fully revised in the *Memoirs* with the copy put through the press by Baily, only reveals one serious alteration of the sense. The original reads :—

> A well-made instrument will thus unavoidably acquire a reputation, not merely among a few eminently skilled observers as at present, but throughout the astronomical world of Britain.

but the last two words have clearly got into the wrong place, and in the revised version are restored to the other leg of the contrast—" a few eminently skilful observers in Britain as at present, but throughout the whole astronomical world."

It is, however, interesting to see what did ultimately happen to one sentence which Sir John had written, and then struck out in MS., but which Baily nevertheless printed. When Sir John got his dozen copies he again struck it out, and again Baily pleaded for it.

> I have ventured, however, to retain the passage in page 5 which you had a second time struck out ; it certainly is too impressive to be lost, and so has thought everyone to whom the passage has been read ; so that if we have done wrong, we must *all* bear the blame equally. It is, however, transposed from its former position to one where it was considered more apposite.

The transposition enables us to identify the passage, which is as follows :—

> Yet it is possible that some bodies, of a nature altogether new, and whose discovery may tend in future to disclose the most important secrets in the system of the universe, may be concealed

under the appearance of very minute single stars no way distin-
guishable from others of a less interesting character, but by the
test of careful and often repeated observations.

We may now declare the secondary purpose with which the
former passage was quoted, and which has prompted also the
relation of these details somewhat more fully than occasion perhaps
seemed to warrant. In recurring to these early days after the lapse
of a century, there can scarcely be a better motive than that of
realising what was in the minds of these pioneers, and seeing what
came of it. As one item alone in their programme, they did not
hesitate to announce it as their ambition to survey the whole sky
by co-operative endeavour down to the minutest star visible in
the best telescopes, and that with the laborious methods of the time !
Two-thirds of a century later, with the immensely powerful aid
of photography at hand, their successors really embarked on this
project, but have found it far beyond their resources. The sky
has indeed been completely photographed, at Harvard and elsewhere, but this is only one step on the way to the scrutiny of each
star—" careful and often repeated." The question forces itself
on our attention whether our pioneers had really counted the cost ;
and we can only reply that, if they had not, they were only com-
mitting the same mistake which their successors, with far better
information, repeated in 1887. They then initiated the project
for the Astrographic Chart, which was to be completed in a dozen
years, though to-day, after nearly three times that period, it is
yet far from accomplishment. Of the enthusiasts who adopted so
great a programme in 1820, probably Sir John Herschel had the
best means of knowing what it involved ; and we may perhaps
read into his attempted deletion of the sentence some misgivings *
whether ambition might not overreach itself. Possibly the
cataloguing of every star might be achieved, by sharing out the
work : but what about " careful and often repeated observations " ?
Perhaps that had better go out ? However, the other enthusiasts
were too many for him and it was ultimately retained.

We see then that the infant Society did not merely " hitch
its waggon to a star," but would be content with nothing less
than the whole universe of stars down to the minutest. Fortunately they were nevertheless men who realised well enough that
whatever their ultimate aims might be, their beginnings must be
eminently sober and practical. They started with the reform of
the *Nautical Almanac* ; and read papers to one another about
micrometers and refraction ; or arranged skeleton forms for

* On 1820 December 19, Sir John writes to Babbage: "Why not proceed to
set on foot that ' regular systematic examination of the heavens ' about which
there is so much said *ad captandum vulgus* in the Address ? "

recording observations—hundreds of which remain blank to this day. But the observational programme was, as already mentioned, only one of the considerations put forward in the Address : others were the collation and publication of observations already made or to be made ; the education of observers ; the determinations of position on our earth ; the improvement of lunar tables ; the establishment of relations with foreign astronomers, who may be elected Associates ; the diffusion of information ; the computation of orbits ; the formation of a library ; and the proposal of prize questions.

Besides the Address, there were " regulations " (which ultimately became our " Bye-Laws ") to be made before the first meeting on February 8. The death of the King on January 29 threw a cloud over this first meeting, so that after reading the Regulations, making some slight changes in them, resolving that they should be printed, as also the Address " with such alterations as [the Committee] may think proper " (wherein we probably catch a reflection of Sir John Herschel's grievances), it was decided to defer any but pressing business to " some future day out of respect to the memory of his late Majesty, whose funeral had not yet taken place." Under the circumstances it was creditable that twenty-one members attended this meeting : and the number of those who had formally joined the new Society was reported as forty-seven. The " future day " was fixed as February 29, when twenty-eight attended, including the Duke of Somerset, who was unanimously elected President. The Vice-Presidents were Colebrooke, Groombridge, Sir William Herschel, and the Astronomer Royal (Pond); Treasurer, Dr. Pearson ; Secretaries, Babbage, Baily, and John Herschel (Foreign) ; and Council, Col. Beaufoy, Capt. Colby, Olinthus Gregory, T. Harrison, D. Moore, E. Troughton ; while A. Baily, D. Moore, and C. Stokes were appointed Trustees. The roll of membership was by that time eighty-three. The meeting concluded with votes of thanks to the Geological Society for the hospitality of their rooms for these early meetings, and to Daniel Moore for his good offices as chairman : and so far all had gone well.

But at the next meeting a blow fell : Sir Joseph Banks, the President of the Royal Society, had induced the Duke of Somerset to decline the Presidency. It seems to have been quite unexpected, for Baily did not write about it to Sir John Herschel until March 11, the day after the meeting, and the Duke's letter is dated March 9, the day before. But apparently Sir Joseph Banks had been at work in various directions. Baily writes :—

A similar attack was made by Sir Jos Banks on the Astronomer Royal, who, if report be true, made a very spirited reply. As a similar, and indeed a more violent, attack was made at the establishment of the Geological Society, and also of the Royal Institution, and which only tended to unite more firmly the original members, we hope that a similar result will also be produced here.

It seems also worthy of record that in Sir John Herschel's Diary there are the following entries :—

> *Sat., Feb.* 26.—Committee of Astronomical. Dined with Mr. Baily.
>
> *Sun., Feb.* 27.—Sir J. Banks.
>
> *Mon., Feb.* 28.—Went in morning to Dr. Pearson's at East Sheen where I spent the day. In morning read over with the Dr. his paper on Micrometers for Astron. Society. Mrs. P. & Mr. & Mrs. Moffat were the party besides the Dr. & myself. In evening after a little music returned to London.
>
> *Tues., Feb.* 29.—Took up quarters at Bedford Place & dined there. My Father and Mother came to dinner. Evening, attended the meeting of the Astronomical Society, at which my Father was voted a Vice-President. Babbage, one of ye English Secretaries, & myself ye Foreign.
>
> *N.B.*—Duke of Somerset, Pres., to whom with the other Secretaries I was introduced by Dr. Pearson. Supped with Babbage.
>
> *Fri., Mar.* 3.—Returned to Slough with my Father & Mother, calling on my uncle B. on the way.
>
> *Mon., Mar.* 13.—Wrote to F. Baily in expression of strong indignation at the part Sir J. Banks has thought proper to take respecting the Astronomical Society in persuading the Duke of Somerset to resign ye Presidentship—also desiring information on the subject of the private observatories on the Continent.

The cryptic entry of February 27 between two notes about the new Society, which was clearly occupying Sir John's thoughts to a considerable extent, can scarcely be interpreted otherwise than that Sir J. Banks must have known of the new Society on February 27, though he delayed action until after February 29. But without further preamble let us inspect the Minutes of the March *meeting of Council* :—

MINUTES OF COUNCIL

Friday, March 10, 1820

The first meeting of the Council took place this day at 3 o'clock in the afternoon ; present : H. T. Colebrooke, Esq. ; S. Groombridge, Esq. ; Rev. Dr. Pearson ; F. Baily, Esq. ; Col. M. Beaufoy ; Capt. T. Colby ; T. Harrison, Esq. ; D. Moore, Esq. ; E. Troughton, Esq. S. Groombridge, Esq., in the chair.

SIR JOHN HERSCHEL
(1792–1871)

To face p. 8.]

A letter (of which the following is a copy) was received from his Grace the Duke of Somerset, viz. :—

"PARK LANE, 9th March 1820.

"To the Council of the Astronomical Society of London.

"Gentlemen,—The gratification I derived from the appointment, with which your Society honoured me at its last meeting, & which was then only qualified by the apprehension of my inadequacy as to fulfilling its duties, has since been obliged to yield entirely to a feeling of a different kind, and one which (I am sorry to say) will no longer allow me to hold that high situation. The professions which terminate the Address of your Society, and some great names which are to be found in the list of its members, had given ample ground for trusting that, as nothing was intended inimical, so, nothing could follow prejudicial, to the interests of an old respectable and chartered body. Its President is however of quite a different opinion, and apprehends the ruin of the Royal Society. To Sir Joseph Banks I have been long & strongly attached, not only by the ties of public regard, but those of private friendship ; and my remaining in a post, which he considers as a hostile position, might be liable to unfavourable comments, & would certainly be very painful to my own feelings. I trust therefore you will not wonder that, under the influence of these impressions, I feel myself obliged to resign the flattering hope of connecting my name with the labours of the Astronomical Society, & that I am under the necessity of withdrawing from the list of its members : & that I am indeed to hope that you will receive this my immediate & sudden resignation & recession with that indulgence which I can only claim on account of the motive which I profess.—I have the honour to be, Gentlemen, your much obliged & obedient serv.

"SOMERSET."

A letter was also received from John Fuller, Esq., stating that he did not mean to belong to the Astronomical Society, because it did not meet with the approbation of Sir Joseph Banks.

The Treasurer reported that A. Baily, Esq. ; F. Baily, Esq. ; Capt. Colby ; D. Moore, Esq. ; J. South, Esq. ; and himself and Mr. Troughton had each paid the sum of Twenty guineas, as a composition for their future annual contributions.

Resolved unanimously—

That the several compositions for the annual contributions which have been, and may hereafter be, received by the Society, shall be, from time to time, invested in the Navy 5 p. cents., in the joint names of the Trustees of this Society for the time being ; as a separate fund.

Resolved unanimously—

That the Capital Stock, created by such investment, shall remain as a permanent fund, the interest only of which shall, if necessary, be appropriated to the current expences of the Society.

Resolved unanimously—

That the Treasurer be required to lay out, from time to time, in public securities, such other sums in his hands, not requisite for the current expences of the Society.

The above minutes were read and admitted to be correct, *April* 14, 1820.

FRANCIS BAILY, Sec.   H. T. COLEBROOKE.

At this distance of time we can afford to regard with equanimity the action of Sir Joseph Banks, and to recall some circumstances which may remove any feeling of bitterness. He had had a long and distinguished reign of forty years over the Royal Society, but it had nearly come to an end. On 1820 May 18, age and illness led him to tender his resignation of the Presidency ; but it was not accepted by the Council, and he accordingly withdrew it. The early R.A.S. Minutes have reminded us of the death of King George III. on January 29. When the Bishop of Carlisle and the Vice-Presidents of the Royal Society waited upon the new monarch with the book of signatures, he took occasion, after inscribing his name as Patron, to congratulate the Society that Sir Joseph Banks should have withdrawn his resignation and continue in office : and doubtless His Majesty represented public opinion. But the continuance was for a few months only, and on the death of the veteran, the new President, Sir Humphry Davy, was most cordial to the new Society. We need not make too much of the fears of an old and enfeebled man : nor need we regard unfavourably the action of his friend the Duke of Somerset in refusing to wound him at such a time. Their relations were apparently closely personal, as we may gather from the Duke's letter itself, but more definitely still on looking up his personal history. According to *The Times* for 1855 August 16, he named his third son Algernon Percy *Banks* Seymour (1813–94), doubtless out of regard for his friend. [The son of this third son ultimately succeeded to the title, following his father after all three sons had inherited in turn.]   The Duke chosen as President was the eleventh (1775–1855), and succeeded to the peerage on his father's death in 1793. He was F.R.S., F.S.A., F.L.S. ; President of the Royal Literary Fund (1801–38) and of the Royal Institution for some years ; " an excellent landlord ; " supported the repeal of the Corn Laws ; and wrote books on the ellipse and circle (1842 and 1850). Our founders seem to have made a thoroughly good choice, and we may well regret that regard for an old friend's feelings, however mistaken they may have been, prevented so worthy a man starting us on our way.

The first impulse of the executive was to find a new President of somewhat similar type. "Application has been made to a nobleman, of nearly equal rank," writes Baily to J. Herschel on March 11, "who is less dependent on the opinion of others, and who is more fitted for the situation." But apparently the reply was unfavourable, for on March 19, Herschel wrote to Babbage asking, "Has the Marquis of Abercorn been suggested? I can't say, Sir X.Y. seems to me a very proper man. Lord Lansdowne may do very well. If we want rank, why should not the Duke of Gordon, one of our members, be applied to? Here is a queer problem of chances which leads to such diabolical series as resist hitherto all my attempts to find their law. A and B toss up— they play double or quits—In n games what is the probability that the stake will attain 2^i × original stake." And so the great and versatile man passes from the problems of the moment to those of more permanent interest. Ultimately it was decided not to elect a new President until the end of the year, and then Sir William Herschel, already a Vice-President as we have seen, accepted the higher Chair, though, on the understanding that he should not be called up for active service. A letter of John Herschel to Babbage shows that a tentative proposal of the same kind had been made by the Council in April 1920, but at that time his father was unwilling to take the office even under the conditions proposed.

When, shortly afterwards, he died full of years and honours, Colebrooke, who had often represented him in the Chair, was elected his successor.

The new Society being now fairly launched, let us see how they began work.

THE MEETINGS.

For accounts of the meetings of the Society nowadays we should go to the *Monthly Notices* or to the more informal records of discussion in *The Observatory* magazine. But neither of these sources of information exist for the early years. The latter is a comparatively recent institution; the *Monthly Notices* extend back only to 1827, and though something can be done to extend them to 1820, as will presently be shown, even then they are but meagre. We are thrown chiefly on the Minutes and a few brief contemporary references; and the general impression created is one of admiration for the hardihood of those who lived through the early meetings. Apparently it was the custom to read the papers pitilessly through, and a long paper might extend over more than one meeting. The *St. James's Chronicle* of 1820 May 13–16, records (in a scrap preserved by Dr. Lee) as follows :—

Astronomical Society.—The first meeting of this Society was held a few days ago at the house of the Geological Society, Bedford

Street, Covent Garden, and was very numerously attended. A paper by the Rev. Dr. Pearson was read on the subject of a new micrometer which he had invented for measuring small distances in the field of a telescope. . . .

At this first ordinary meeting, held at eight o'clock on March 10, the Minutes tell us that twenty-eight members were present, Stephen Groombridge being in the Chair. Three new members were proposed for election (being numbers 84, 85, and 86), a score of books were presented by Brisbane, Stokes, Hutton, Colby, and Baily, and then Dr. Pearson's paper was read " On the doubly-refracting property of rock-chrystal, considered as a principle of micrometrical measurements when applied to a telescope." The summary of its contents (two pages of MS. minutes) ends with the statement, " The practical application of the micrometer thus constructed the author proposes to communicate at a subsequent meeting of the Society."

At the meeting of April 14, Colebrooke was in the Chair; 22 were present; 13 new members were proposed; and Dr. Pearson gave his promised description of the construction and use of his micrometer, producing actual instruments and measures made with them.

On May 12, Groombridge in the Chair, 24 members *and one visitor* present; 5 new members were proposed; the 3 proposed in March were elected (thus initiating the adopted practice of election two meetings after proposal). Dr. Brinkley, Professor at Dublin, one of the five proposed, explained that his tardiness in coming forward was due to accidental delay in his receiving the original circular, which he much regretted; news of the proposal for an observatory at Cambridge was announced; and Mr. James South read a paper on double stars.

On June 9 (Colebrooke in the Chair), no list of members present is given. Eight new members and one Associate (Biot) were proposed; the 13 proposed in April were all elected, and 2 of the 3 elected in May were formally admitted (there was, however, as yet, no book for them to sign : that came later). Captain Basil Hall announced that he was sailing in a frigate to the south of Cape Horn, and would be happy to receive instructions for nautical observations likely to be of value. F. Baily read a paper " upon a method of fixing a transit instrument exactly in the meridian," and Sir H. Englefield addressed the meeting *orally*. From the fact that he apologised for doing so, owing to " his present inability to *write* on any subject " we may infer the rigidity of the rule that communications must be in writing. He drew attention to some old observations which might refer to comets, or perhaps satellites to Venus, and suggested further search.

In November (Colebrooke in the Chair; 9 members and 7 Associates proposed, 6 members elected, 3 admitted) it was announced that a Committee had handed a paper of instructions to Captain Basil Hall (in 1822 the same instructions were handed to Captain Owen, who was going to the east coast of Africa). Attention was drawn to the favourable position of the moon for occulting the Pleiades; two books were presented, and also a bronze medal of Copernicus presented by Peter Slawinski, one of the 14 who originally met on January 10; Groombridge read a paper on star-reductions; and Gauss on a new meridian circle at Göttingen.

In December (Dr. Pearson, Treasurer, in the Chair) 2 members and 4 Associates were proposed, 8 members and 1 Associate (Biot) elected and 4 members admitted. Four books were presented; and papers read by Groombridge (ephemeris of Vesta) and Baily (solar eclipse of 1820 September 7). Troughton began his paper on the Repeating Circle and Altazimuth, but neither this meeting nor that in January sufficed for the complete reading of it, and it occupied also the whole of the March meeting (February being devoted to the annual meeting, according to the practice still followed). But the story was again taken up in April by George Dollond, who gave *his* views about the Repeating Circle: so that four consecutive ordinary meetings were largely devoted to this instrumental description, the last being relieved only by some observations of comet Pons-Nicollet sent by Olbers.

These brief notes of the first year's work will suffice to illustrate its nature. It is needless to say more of the papers themselves, which are all printed fully in the first volume of the *Mem. R.A.S.* There was not much to attract the populace: but the men who formed the backbone of the Society were made of stern stuff, and often when the ordinary meeting was over, sat down again to carry on the work which they had failed to complete at Council in the afternoon. Council met at three o'clock, and probably continued its labours until the hour for the dinner at which we know (from other records) that the chief councillors generally assembled. From this they came back to the eight o'clock meeting, with its frequent appendix of an adjourned Council. In one burst of enthusiasm, Sir John Herschel moved that the meetings of the Society should be held *twice* in the month during the session, but after some discussion thereon he withdrew the motion (Council Minutes, 1821 April). But the Council had occasionally to meet twice. We may rightly infer that the work of the Council in those early days was by no means the least important part of the work of the new Society. The irreverent may perhaps hint at the attraction of the Club dinner, but it is a significant fact (which may surely be borrowed without impropriety from the Club Annals), that on occasions when there was a dinner without a meeting of the Society and the Council, the attendance was apt to be small. This fact is explicitly deplored in the Annals afore-said, and from the point of view of the Club, which called these meetings specially for the regulation of its own affairs, the regret is intelligible. But in compensation we get the knowledge that what really attracted these earnest men and brought them together was the work : the courageous endeavour to do something not only for the Society itself but for Astronomy generally. Round the Council table they discussed how to stimulate astronomical research by offering prizes ; how to obtain better object glasses for telescopes (it will be remembered that the Herschels worked with mirrors) ; how to make astronomical tables ; how to arrange convenient forms of reduction for observations (we are accustomed to associate such forms with Greenwich, and especially with Airy ; but Pearson and Baily used them long before Airy) ; how to mea-sure the length of the second's pendulum ; and how to improve the *Nautical Almanac*. When we remember that they had also to start the new Society from the cradle, to build up its funds and its library, to arrange for the printing of its *Memoirs,* even to find a home for it as mentioned below, we see that there was plenty of work for the Council ; and it does not need much imagination to trace the origin of the earnest spirit which fortunately still animates it to those early days when there was so much to be done and so little to look back upon as achieved.

A few instances will suffice to illustrate the history of those early years. On Thursday, 1820 November 30, the Council met at 10 a.m. at Baily's house to consider a request (signed by South, Fallowes, G. Dollond, P. Kelly, B. Donkin) for accurate tables of the 45 Greenwich stars for 1822, 1823, 1824. Now that we have a "clock star" every few minutes, we may well admire the restraint of those who pleaded for one every half-hour. The actual request was not pressed when the Council promised to do its best. According to a report made a week later the main obstacle was the indolent Board of Longitude. In the *N.A.* for 1822 there is, indeed, a list of the 45 stars, but ephemerides are only given for 24 of them, so that there were such gaps as $2^h 28^m$ (a Arietis to Aldebaran) and $3^h 18^m$ (Regulus to Spica) during which an observer could not conveniently find his clock error. This state of things continued for several years, the *N.A.* for 1826 showing no improvement. Perhaps we may reproduce (from the *N.A.* for 1822) the names of the Board of Longitude. From their laxity we should rather expect to find them officials innocent of Astronomy, but they include some very respectable names, especially that of the Secretary :—

BOARD OF LONGITUDE, 1820

J. W. Croker, S.A.	S. P. Rigaud.
Jos. Banks, P.R.S.	Isaac Milner.
Davies Gilbert.	Samuel Vince.
Rob. Woodhouse.	W. Lax.
John Pond, Ast.R.	W. H. Wollaston.
A. Robertson.	W. Mudge.

THOS. YOUNG, Secretary.

With the rebuff encountered by the new Society in our minds, we read these names with surprise, and may reflect on the justice of the saying that a Board or Committee is apt to combine not the wisdom but the folly of the members, or shall we modify it by substituting " not the activity but the inertia."

It is a pleasure to contrast the conduct of the R.A.S. Council, which set about calculating and printing the requisite ephemerides for 1823. Such work was followed up until it resulted in *Baily's Catalogue* of nearly 3000 fixed stars (*Mem. R.A.S.*, 2, Appendix) with the " star constants " a, b, c, d, etc., and day constants for every tenth day for the years 1826–30, a really magnificent achievement for a Society (indeed, almost for an individual member of that Society) in the face of official laxity and discouragement. This method of computing " star corrections " is now so familiar that we find it difficult to imagine the state of affairs before Baily introduced it into England and ultimately into the *Nautical Almanac* for 1834, which was " constructed in strict conformity with the recommendations of the ASTRONOMICAL SOCIETY of London "—a great triumph for the Society, of which more will be said in the next chapter. Baily did not invent the method, he took it from Bessel and Schumacher, with a modification of his own for which he offers the following reason * :—

It may be proper here to state that the values denoted in the present tables by A, B, C, D, are denoted by M. SCHUMACHER C, D, A, B respectively. But, in the choice of characters to represent given quantities, it is desirable that we should, as much as possible, make them serve the purpose of an *artificial memory*. It is on this account that I have made A, B represent the quantity by which the A Berration is determined ; C the quantity by which the preCession is determined ; and D the quantity by which the Deviation, or (as it is now more generally called) the nutation, is determined.

The reason seems a good one, and it is perhaps a pity that the

* *Mem. R.A.S.*, 2, xxx, footnote.

stolid obstinacy of the Germans should have triumphed over it. In the *N.A.* for 1916 the Superintendent has abandoned Baily's notation for the German.

Turning to different matters, in 1823 February it was reported to the Council that Tulley had completed a 2-in. O.G. from glass by Guinand of Neufchatel, which Dollond had found satisfactory. [The spelling of such names in the Minutes varies considerably.] A Committee of Gilbert, Herschel, and Pearson was thereupon authorised to purchase similar glass " on account of the Society " to an amount not exceeding £100 : but it was reported in April that the maker had no adequate supply of the glass. Tulley's telescope was purchased by Baily for 14 guineas, after others had declined it (May 9). In November a further supply of glass from Guinand was reported : viz., 3 pieces of flint glass, 2 amorphous, and one as a disc for a 7¼-in. O.G. This disc was put into the hands of Tulley and Dollond, and ultimately Tulley fashioned an O.G. of nearly 7 inches aperture which Dr. Pearson purchased, giving £200 to Tulley, and paying £20, 16s. 6d. + 700 francs for the glass to Guinand. A report on the whole transaction is printed in *Mem. R.A.S.*, **2,** 507 ; but neither this brief summary of facts which to-day seem unimportant nor the report mentioned can convey an adequate idea of the time and thought spent by the Council, at many meetings, in this attempt to obtain better re-fracting telescopes. They were grievously disappointed at Tulley's charge of £200, and told him so, pointing out what a discourage-ment it was to further work. It was only the kindly generosity of Dr. Pearson which smoothed over an awkward situation.

Before the Beginning

Let us now, before following the history of our Society further, turn back to some circumstances attending its inception. In the *Memoirs* of Augustus De Morgan (sect. iii. p. 41) the following remarks of Sir John Herschel are quoted :—

> The end of the eighteenth and beginning of the nineteenth centuries were remarkable for the small amount of scientific move-ment going on in this country, especially in its more exact depart-ments. . . . Mathematics were at the last gasp, and Astronomy nearly so—I mean in those members of its frame which depend upon precise measurement and systematic calculation. The chilling torpor of routine had begun to spread itself over all those branches of science which wanted the excitement of experimental research.

The foundation of our Society was thus associated with an awakening from this deplorable state of affairs. We must be careful to note the qualifying phrase with regard to Astronomy ;

it was on the mathematical side that it was defective : on the observational side the period mentioned was precisely that of the immortal work of Sir John's own father, and was not likely to be overlooked.   We can the better understand why the early activities of our Society were chiefly concerned with the stimulation of progress where it was most needed.   But we were fortunately not the only new Society, nor even the first.   The facts may be briefly recalled by quoting from an article in the *Quarterly Review* for June 1826 :—

> From the institution of the Royal Society in 1663, to the year 1788, when the Linnean was founded, no subdivision of scientific labour was attempted in our metropolis. The Royal Society continued, without assistance, to embrace within its aim the cultivation of every department of natural philosophy ; but a further subdivision of labour, as inseparable a consequence of the progress of the sciences as of the arts, was at length effected with the concurrence and co-operation of the Royal Society itself ; and the prosecution of the studies of zoology and botany in all their details was the chief object of the institution of the Linnean Society, which received a royal charter in 1802, and has now published fourteen volumes of *Transactions*, containing a variety of most valuable memoirs.

The Royal Institution, the next in order of date, was founded in 1799, and the College of Surgeons in 1800.

> The Horticultural Society, established in 1804, although designed to promote luxury rather than science, must not be omitted here. . . . The London Institution, " for the advancement of Literature and the diffusion of Useful Knowledge," was founded in 1805 and chartered in 1807. . . . The Geological Society of London, established in 1807 and chartered in 1825, has been eminently successful in giving the impulse to the study of geology in Great Britain. . . . The institution of the Astronomical Society of London in 1821 (*sic*) was actively promoted by many of the most distinguished fellows of the Royal Society. Besides the excellent volume of *Transactions* already published, we have pleasure in being able to state other important benefits which have resulted from their efforts. A valuable set of tables for reducing the observed to the true places of the stars is preparing at the expense of the Society, including above 3000 stars, and comprehending all known to those of the fifth magnitude inclusive, and all the most useful of the sixth and seventh.

In addition to this the reviewer mentions the machine called Babbage's Calculating Machine, which had already secured Government encouragement, and continues :—

> After this brief enumeration of the chief scientific institutions of the metropolis, which the reader cannot peruse without being

struck with their recent increase, we hasten to consider the rise
and progress of similar institutions in the provinces.

He mentions first the Observatories at Oxford (Radcliffe),
Dublin, Armagh, Cambridge, and the private Observatories of Mr.
South and Mr. Herschel, commenting adversely and emphatically
on the fact that " no public observatory where observations are
regularly made exists at present in Scotland." Mention is also
made of the Ashmolean Society of Oxford, the Literary and
Philosophical Society of Manchester (1781), the Royal Geological
Society of Cornwall (1814), the Liverpool Royal Institution (1814),
the Cambridge Philosophical Society (1819), the Bristol Institution (1820), the Yorkshire Philosophical Society, and " many other
institutions in our provinces, such as those of Newcastle, Bath,
Leeds, and Exeter."

OUR FOUNDERS

It seems further desirable to give here a few words about some
of the men who founded the Society. About a few of them information is already available in plenty, as, for instance, in the case
of Sir William Herschel, the first nominal President of the Society,
and there is no need to say more here ; but the case is somewhat
different with the second President—the first who actually filled
the Chair, H. T. Colebrooke. His name may be quite unfamiliar
to most astronomers, and yet he was a very remarkable man. He
died in 1837 after some years of suffering both bodily and mental,
and our *Monthly Notices* of the time (**4,** 108) give little beyond a
reference to a short Memoir in the *Annual Report of the Royal
Society*. But the essay by Max Müller, which appeared in the
Edinburgh Review for October 1872, and was reprinted in *Chips
from a German Workshop*, and in the *Biographical Essays* (Longmans, 1884), enables us to form some estimate of the intellectual
stature of Colebrooke. Max Müller calls him the " Founder and
father of true Sanskrit scholarship in Europe," and remarks with
some bitterness that if he had lived in Germany his name would
have been written in letters of gold on the walls of academies ; but
that in England, though we may hear the popular name of Sir
William Jones, we hear not one word of the infinitely more important achievements of Colebrooke.

To show that this is not a careless comparison, he returns to
it at the end of his essay, and deliberately declares that, " as
Sanskrit scholars, Sir William Jones and Colebrooke cannot be
compared. Sir William had explored a few fields only, Colebrooke
had surveyed almost the whole domain of Sanskrit literature."

Those interested will, from this reference and with this guiding estimate, be able to follow up this line of thought for themselves. What specially interests us is the beginning of this interest in Sanskrit, which was from the first scientific rather than literary. Colebrooke's love of mathematics and astronomy made him anxious to find out what the Brahmans had achieved in this branch of knowledge, and Max Müller draws attention to the surprising correctness of his first letter to his father on the four modes of reckoning time adopted by Hindu astronomers. " In stating the rule for finding the planets which preside over the day called *Horâ*, he was the first to point out the palpable coincidence between that expression and our name for the twenty-fourth part of the day." *   But that his literary enthusiasm was at this time not very great is clear from his reference to other scholars, and his opinion that all to be expected from Sanskrit was that a few dry facts might possibly reward the literary drudge.   He himself took up the study and left it again in despair several times, and in 1793 wrote that " no historical light can be expected from Sanskrit literature ; but it may, nevertheless, be curious, if not useful, to publish such of their legends as seem to resemble others known to European mythology," at which Max Müller exclaims : " The first glimmering of comparative mythology in 1793 ! "   Even then his studies were guided by a practical rather than by a literary motive. He was keenly interested, for instance, in the agriculture of the Hindus, and possibly not only the Astronomical Society and the Asiatic Society might reckon him as a pioneer, but also those who study the history of agriculture.   The Asiatic Society he founded in 1822, though he refused to become the first President.   We may regard it therefore as specially significant that he occupied our own Chair at about the same date.   He had spent thirty-three years in India, having arrived there in 1783 when only seventeen years of age, and left it in 1815 at the age of fifty.   His essays were collected by his son, who added a brief life of his father, and it was the appearance of a new edition of these two volumes that gave occasion for Max Müller's essay.   The portrait of him which hangs in our meeting-room is from a painting in the possession of the family, and was kindly presented to us.

One point of detail may be mentioned.   On 1821 June 8 the Council, who had to settle the type for the *Memoirs*, resolved to adopt the same as that of Colebrooke's *Indian Algebra*.

---

* In Colebrooke's *Miscellaneous Essays*, vol. ii., there are two lengthy papers (reprinted from the *Asiatic Researches*), "On the Indian and Arabian Divisions of the Zodiac" and "On the Notion of the Hindu Astronomers concerning the Precession of the Equinoxes and Motions of the Planets." Also an essay "On the Algebra of the Hindus," reprinted from Colebrooke's translation of Brahmegupta's *Algebra*.

Before Colebrooke, however, before Sir W. Herschel, before even the Duke of Somerset, we had as effective President Mr. Daniel Moore, who was specially thanked for taking the Chair at the preliminary meetings. What manner of man was he? He died in 1828 before it had become the custom to give many biographical details of our deceased Fellows. We read in the Council Report of " our amiable and excellent trustee Mr. Daniel Moore, whose loss will be felt far beyond the limits of this body by many, as the privation of a benefactor, in whose ears the calls of distress never sounded in vain," and that is all : but it is much. A casual reference in one of Sir John Herschel's letters gives us almost the same picture. He is " our friend Moore, whose money burns in his pocket," and who might come to the rescue " if the low state of [the Society's] funds be talked of." It is but a glimpse we get, but a very pleasant glimpse.

Such were the men who took the Chair at the early meetings, either actually or nominally. But there is no question that for real initiative the Society owes almost everything to two men, the Rev. William Pearson and Francis Baily. Probably the combination of the two was really necessary. The dreamer Pearson had long had the project vaguely in mind, but required the help of Baily, a man of affairs, to put it into practical shape. The incidence of Baily can be traced in his Appendix * to a *Memoir on a new and certain Method of ascertaining the Figure of the Earth by means of Occultations of the Fixed Stars.* By A. Cagnoli. With Notes and an Appendix by Francis Baily. London, 1819. 8vo.

He therein (p. 29) strongly urges the formation of an ASTRONOMICAL SOCIETY, with a library and a collection of observations, referring to Pingré's *Annales Célestes* : and that the scheme took shape within a year strongly suggests that this new and vigorous influence was the determining cause. But Dr. Pearson had had the idea as early as 1812, and it was he who ultimately assembled those interested at a friendly dinner † in order to hatch out the project. The facts are given in two letters which were printed by De Morgan, and are bound up with some copies of the *Monthly Notices* (26), but not with all. It seems, therefore, desirable to reproduce them here : with the comment that (in spite of the

* I am indebted to Dr. Dreyer for this reference, which he first found in the library of the Radcliffe Observatory, Oxford. It is catalogued in the R.A.S. and Crawford libraries, but under Cagnoli, and in the former case with no cross-reference.

† Until seeing the entry in the Diary of Sir John Herschel, I had always supposed that this dinner was on a date before January 12, and probably at Dr. Pearson's house ; but the facts seem consistent with the dinner being that at Freemason's Tavern, immediately preceding (or following) the meeting of January 12.

statement made in *M.N.*, **8,** 73, that the dinner was at East Sheen) the dinner was probably simply that at Freemason's Tavern as mentioned in Sir John Herschel's Diary (see p. 1).

LETTER FROM A. DE MORGAN, ESQ., TO THE PRESIDENT, ON THE FOUNDATION OF THE SOCIETY

" On looking over old papers, I find copies—whence obtained, I forget—of two letters connected with the foundation of the Society, in the handwriting of Mr. B. Smith, who was Dr. Lee's secretary. With them I found a letter from Dr. Lce (September 19, 1857) in answer to my inquiries. It appears that the originals had then been, for many years, in possession of Captain Smyth, who entertained, from 1830 to 1834, or thereabouts, the intention of writing on the foundation of the Society. The copies are addressed to Mr. Sheepshanks in the handwriting of Captain Smyth, with the postmark 'Bedford, May 24, 1834.' My impression is that Mr. Sheepshanks handed them to Mr. Baily, among whose papers I should have been sure to have found them. In this I am somewhat confirmed by observing that Mr. Sheepshanks, in his obituary notice of Dr. Pearson (Annual Report, 1848), shows only a general recollection of the first letter, and none at all of the second. As Admiral Smyth and Dr. Lee are now gone, and probably no one but myself knows of the letters, I think it right to put their contents on record.

" The first is from Dr. Patrick Kelly (the author of the *Cambist*) to Dr. Pearson, December 12, 1812. He says: ' It [a meeting of schoolmasters] may be also a very auspicious time for us to lay some foundation for your suggestion respecting an *Astronomical Society*. I have mentioned it to two or three scientific gentlemen, who all approved very much of the idea ; and one in particular, Mr. [Peter] Nicholson, thinks that under good management it might become of great importance to science.' In a postscript Dr. Kelly adds : ' If the *Astronomical Society* should ever become great, you must not forget that you are the *Father of it*. There are several eminent societies in town possessing inferior objects.' It thus appears that Dr. Pearson had formed the plan by 1812 and was endeavouring to promote the formation.

" Mr. Sheepshanks mentions, as a rumour, that the meeting of January 12, 1820, at which the Society came into existence, was resolved upon at a dinner given by Dr. Pearson. The second letter fixes this rumour as a fact. It is from Mr. (Sir James) South to Dr. Pearson, December 13, 1819, giving permission to add the writer's name to a list then in collection, and accepting an invitation to dinner ; the date of the symposium is not given.

It thus appears that Dr. Pearson kept the plan in his head, where it lived through his transformation from a thriving London schoolmaster into a country rector and magistrate, that he got together a number of astronomers to join him, and lubricated the business, to use Sam Johnson's phrase, by a dinner. I may be permitted so much reference to our age and country as will appear in a slight alteration of Molière's text :—

> " Le véritable fondateur
> Est le fondateur où l'on dine.

" Francis Baily (1819) gave in print a recommendation that such a Society should be formed. Sir J. Herschel, when he wrote his life of Baily, was not aware that Dr. Pearson had been agitating the plan for seven years. Dr. Pearson, who finally left London in 1821, could not have been, what Baily was from the very first, the guide and stay of the Society, an institution which many might have founded, but few could have nursed. If the word be plural both were founders ; but so far as it can be used in the singular it applies only to Dr. Pearson.

" It must be remembered that in 1820, Dr. Pearson stood in a position which the Society gradually altered by raising others to his level. He had that knowledge, which his work of 1824 so amply shows, coupled with great industry and zeal, and a remarkable collection of instruments. His standing in society was good, and his character high. To us Baily is what he made himself in making the Society : but in 1820, though Baily was well above the horizon, Pearson was on the meridian.

" It is to be remembered that we are not to assume that we know of all Dr. Pearson's exertions in this matter. Action in 1812 and action in 1819, proved by record, may lead to more than surmise of something like continuous effort through all the intervening period. My floating recollections of what people said in 1830 tend to strengthen the conclusion that Dr. Pearson never lost sight of his favourite project."

Dr. Pearson died in 1847 and the Council Report (*M.N.*, **8,** 69) contains a notice of his life over which much pains was clearly taken, but which ends apologetically for its meagreness. He was born in 1767 at Whitbeck, Cumberland, and resided for some time at Lincoln, where he constructed a portable clock which showed the age of the moon. De Morgan wrote to John Herschel in 1867 : " I have just found out that Dr. Pearson began life as a junior partner in Sketchley & Pearson, who kept a school at Fulham for boys from four to ten. Here he had been for some years in 1800. I picked up a sensibly written prospectus—they said plan then—

of this establishment. I make out that he was not a graduate of Oxford or Cambridge " (*Mem. of A. De M.*, 372). In 1810 he " became the proprietor of a celebrated establishment at Temple Grove, East Sheen, where many of the nobility and gentry received their preparatory education. Here he built an Observatory and furnished it with instruments." In 1817 he was made rector of South Kilworth, Leicestershire, but he continued to reside at East Sheen until 1821, and was thus able to play his leading part in the foundation of our Society. He removed in 1821 to South Kilworth for the rest of his life, and erected there a considerable Observatory, employing a permanent assistant. In 1824 and 1829 he published the two volumes of his *Treatise on Practical Astronomy*, a monumental work of which he ultimately presented the unbound sheets to the Society.

Since so little is known about our Founder, whatever more can be gathered is of value. From two directions sidelights have recently been thrown on his doings. The first is from the Reports of the Yorkshire Philosophical Society, already mentioned (see p. 18) among those founded about the same time as our own Society, and specially distinguished by the part it played in the foundation of the British Association at York in 1831. Dr. Pearson was one of those who attended this first meeting (of the B. A.), and he took occasion to present to the Yorkshire Philosophical Society his *Treatise on Practical Astronomy*. The Society thereupon elected him as Honorary Member. He replied by offering to present some valuable astronomical instruments (later he specified a clock, a telescope, and a transit instrument) if the Society would build an Observatory to house them. We may give verbatim an extract from the Reports for 1832, 1833, and 1857 :—

> Dr. Pearson has given fifty copies of his Tables for the Reduction of Astronomical Observations. This munificent patron of Astronomy will contemplate with satisfaction the Observatory which is now rising to receive his instruments and employ his useful tables. The Committee appointed for this object have been scrupulously attentive to the main point of a solid foundation and an immovable basis for the instruments ; they have made provision for a large transit and a circular instrument, and by placing the revolving telescope on a separate foundation, believe that they shall at once secure accurate observations for time and position, and allow, on suitable occasions, more popular views of the heavenly phenomena.
>
> 1833. The Observatory has been put into active operation and the labours of the Committee for Science have been assiduous and productive.
>
> 1857. Important improvements have been made in the Observatory. The object glass of the telescope (4 in. in aperture) presented by the Rev. Dr. Pearson, for which the Observatory was built,

was found upon examination to have been altered so as to be out of form, and a new one having been supplied by the liberality of one of the Curators of the Observatory, William Gray, Esq., it has been remounted in a new tube provided with eye-pieces, and the instrument has thus been rendered very perfect.

The portable transit by Jones has disappeared, and we see that the equatorial was replaced in 1857 : but the sidereal clock by Barraud is still in the Observatory and keeps fair time, as the present observer kindly informs me. He adds that Dr. Pearson also presented the conical roof of the Observatory, which had served as the roof of a summer-house in his rectory garden at South Kilworth, and was constructed under the direction of the celebrated Smeaton. It may perhaps be added that there had been an Observatory in York before that due to Dr. Pearson's stimulus, viz., that of Edward Pigott, who gave its longitude (from occultations) to Maskelyne in 1787 ($4^m 25^s$ W.). It was from this Observatory that Goodricke observed Algol and δ Cephei in 1782 ; and Goodricke's papers are still preserved in (Dr. Pearson's) Observatory.

The school at East Sheen started by Dr. Pearson did not by any means close when he went to South Kilworth. Our Fellow, Admiral Sir H. Purey-Cust, was a scholar 1866–70, and has kindly supplied some picturesque details about it, partly from the Temple Grove Register (by H. W. Waterfield, 1905). Temple Grove, formerly called Sheen Grove, was built in 1610. It has been generally supposed that Sir William Temple lived there, and that with him, as secretary, lived Jonathan Swift, better known as Dean Swift. Here Swift became acquainted with the beautiful and accomplished Stella, born at this place and the daughter of Sir William Temple's steward. (The same claim is, however, made for Moor Park, another residence of Sir William Temple.) " The property descended to the first Lord Palmerston, who subsequently sold it to Sir John Barnard. In or about 1810 it was bought by Dr. William Pearson, who came from Parson's Green and apparently brought a school with him. On part of the estate he built the Observatory, where an inscription round the central pillar runs as follows :—To the memory of the Right Hon. Spencer Perceval, who was cruelly murdered on the 11th of May 1812, on which day this edifice was also founded, the subjacent pillar is dedicated by his grateful friend W. Pearson."

Both Dr. Pearson (headmaster, 1810–17) and Dr. Pinckney (headmaster, 1817–1835) lived in the Observatory after retiring from the headmastership.

There is a tradition that Benjamin Disraeli was at the school, based on a passage in *Coningsby* describing how the hero was sent to a " fashionable school " (cp. the extract from the Council

Report above) preparatory to Eton, where " he found about 200 youths of noble families and connections, lodged in a magnificent villa, that had once been the retreat of a Minister, superintended by a sycophantic doctor of divinity, already well beneficed and not despairing of a bishopric by favouring the children of the great nobles." As Disraeli was born in 1804, his schooldays would have been about Dr. Pearson's time ; but his biographies mention Blackheath and Walthamstow as his early schools. So that we feel sure that the above rather unpleasant portrait has nothing to do with our Founder, in spite of the following local allusion :—

> " Mr. Rigby was so clever that he contrived always to quarter Coningsby on the father of one of his school-fellows, for Mr. Rigby knew all his school-fellows and all their fathers. Mr. Rigby also called to see him, not unfrequently would give him a dinner at the Star and Garter, or even have him up to town for a week to Whitehall."

If the *Star and Garter* is to be taken literally it certainly points to East Sheen : but it may surely be a substitution for some other famous dining place such as *The Ship* at Greenwich. Disraeli was at school at Blackheath, and by an odd coincidence there was a school there also associated with the name of Spencer Perceval— afterwards divided into two houses, Spencer House and Perceval House.

In later years there was at Temple Grove a pupil whose name (disguised) is even better known than that of Disraeli. In *Tracks of a Rolling Stone* (1905) Mr. William Coke describes Temple Grove as he knew it in 1837. He gave his name to the Billy Coke or billycock hat, otherwise known from its maker, Mr. Bowler. Lord Selborne and Lord Grey were also at the school, the former as a contemporary of Admiral Purey-Cust. Another of our Fellows, Colonel A. C. Bigg-Wither, was there in 1853-55.

Certainly the works of Dr. Pearson, as we know them, do not savour of a " sycophantic doctor of divinity." His generosity seems to have been as great as his assiduity in labours, which many men would find distasteful. It is no light matter to produce a volume of astronomical tables. It is curious how this side of astronomy seems to have fascinated our pioneers : probably it was the link between Pearson and Baily.

We find ample evidence in the history of the early years of the new Society that its prime motive was " precise measurement and systematic calculation." It might have been supposed that the more picturesque work of its first actual President, Sir William Herschel, would inspire the active members to follow him, at however respectful a distance, in examining nebulæ, stellar clusters, and other objects of special interest. Instead of this Francis Baily led them on a campaign of meridian observation, star-corrections, and improvement of the *Nautical Almanac*. Many of our present Fellows would associate these ideas with the " chilling torpor of routine," which Sir John Herschel deplored, and from which (according to his implication rather than his direct statement) the new Society offered a means of escape. But it must be remembered that meridian work has been converted into routine by the century of unremitting toil which stands between us and our Founders—toil which they did much to stimulate, and which has provided us with the luxuries of accurate star-places and star-movements which they were without, and which they enthusiastically set out to acquire. The actual acquisition to-day is, of course, largely due to the work of national observatories ; but the conspicuous figures of Stephen Groombridge and Francis Baily remind us of the important part played by our Society in its early years. Flamsteed and Bradley had started such work nearly a century earlier still, but had been indifferently supported. In 1820 a new impulse was given by our own Fellows of a value scarcely to be overestimated : for even the vigour of Airy might have been less effective but for the support of the Royal Astronomical Society.

As this close connection of our Society (and especially of Baily) with accurate calculations is a really fundamental issue, a few further remarks upon it may be pardoned. Let us look at the following utterance of De Morgan in 1854 :—

> It appears to me that the Royal Society, during the present century, has shown great want of power to appreciate improvements in calculation of results ; and I am afraid I must add, that the University to which I owe my own education has been one cause of this exhibition. I think that for fifty years there was a growing tendency at Cambridge to neglect, in teaching, all that follows the resulting formula or the final equation ; though I suspect that this tendency has passed its culminating point.

These words are quoted from an article in the *Assurance Magazine* * on George Barrett and Francis Baily. The former was a self-taught calculator (1752–1821), who had produced some valuable Life Annuity Tables by a new method of his own. Baily entered into correspondence with him, realised the great value of his tables and method, and endeavoured to get them published. One of his attempts was to submit a paper to the Royal Society in 1811 ; but the paper was rejected. De Morgan comments :—

* A copy was bound up by De Morgan with his volume of Baily's *Journal*, now in the R.A.S. library.

J. J. WATERSTON
(1811–1883)

To face p. 26.]

The rejection of Baily's paper on Barrett's method by the Royal Society is one of those unfortunate instances which create a fear lest there should be other communications, as valuable, which have also been rejected, but have never found such a champion as Baily. It is usual to attribute this rejection to the late William Morgan, who was at that time a member of the Council. . . . But it must not be forgotten that the celebrated Thomas Young, an acute writer on annuities, was also on the Council, and as probably on the Committee. Baily . . . was afterwards, as it happened, in open opposition to Young on the question of the *Nautical Almanac.*

This paragraph suggests several reflections. Firstly, we cannot but think of Waterston's paper on the Kinetic Theory of Gases, rejected by the Royal Society in 1845, but subsequently rescued and printed by Lord Rayleigh (*Phil. Trans.* A, 1892, **183**, 1). He remarks that " the memoir marks an immense advance in the direction of the now generally received theory. The omission to publish it at the time was a misfortune which probably retarded the development of the subject by ten or fifteen years. . . . It is difficult to put oneself in imagination into the position of the reader of 1845, . . . but it is startling to find a referee expressing the opinion that the paper is nothing but nonsense, unfit even for reading before the Society." It is almost equally startling to read the following considered judgment of Lord Rayleigh himself, which gives all scientific societies food for thought :—

The history of this paper suggests that highly speculative investigations, especially by an unknown author, are best brought before the world through some other channel than a scientific society, which naturally hesitates to admit into its printed records matter of uncertain value. Perhaps one may go further, and say that a young author who believes himself capable of great things would usually do well to secure the favourable recognition of the scientific world by work whose scope is limited, and whose value is easily judged, before embarking upon higher flights.

We are strongly tempted to hope that this judgment may not be sound. As regards Waterston himself, he had, to some extent, adopted just the procedure recommended by sending to our own Society (in 1844 June) a short note on a graphical method by which, with ten to fifteen minutes work, an occultation could be predicted within one minute ; and in the January following some good observations of the comet. But, of course, the Royal Society knew nothing of this. The figure of Waterston is so tragically interesting that a brief reference to his later papers may be excused. In 1858 he communicated to us some " Thoughts on the Formation of the Tail of a Comet," which are surprisingly suggestive of modern views of light pressure. He shows himself aware of the difficulties in converting a vibratory movement into a translatory, but he is impressed with the fact that *if* the whole heating power of the sun's rays could be converted into a centrifugal force, the acceleration would be 800 miles per sec. per sec. ; and he is tempted to think that even a small fraction of this might serve. He draws careful distinction between the case of a large body like the earth and a "single and free molecule." In the following year he considered the development of heat in the sun from bombardment by meteors, but had not good enough data to make suggestions of modern value. But he set about improving the data for radiation by experiments of his own, though they still led him to a potential temperature of about 12 million degrees for the solar surface.

There our printed records about him practically stop. But, alas ! there are others in our archives and minutes. At the end of the Council Minutes for 1879 May 9, " A letter was read from Mr. J. J. Waterston, asking that his name might be removed from the list of Fellows of the Society. His request was acceded to." Seeing that he had " compounded " for his annual subscriptions (on 1852 January 9) the step is a remarkable one and led to further search. In 1878 May and June two papers were received from Waterston, and are duly recorded in the printed list (*M.N.*, **39**, 298–9), " On the Heat of the Stars, J. J. Waterston," " On a Solar Thermometer Couple to Measure the radiant Force of Daylight, J. J. Waterston." But they were not allowed to be read or printed. They even appear to have been returned to the author, for they are the only absentees from the bundle of that year in our archives. And their sequel and undoubted consequence was the above withdrawal. These rebuffs undoubtedly prayed on his mind. From occasional remarks his family knew that he had been badly treated by scientific societies, but he never stated his grievances explicitly. He was one of the kindliest of men, and a great lover of children, by whom he was beloved in Edinburgh, where he lived (after his return from India) in Gayfield Square, off Leith Walk. One day in 1883 he went out for a walk and did not return. He used often to go to Leith pier, and it was believed that he had by some accident fallen into the water ; but his body was never found.

It is of small value to look back over a hundred years if we may not pause occasionally to reflect on some of the lessons suggested in the retrospect ; and though this matter of the relationship of scientific societies to their members is something of a digression, it does touch on one of the factors which seems to have been important as a stimulus to Baily, and therefore to our Society ; and it is tragic that we should have nevertheless some memories of a kind he would specially have deplored.

The second point suggested by De Morgan's words is this. We probably follow the bent of his own thoughts in presuming that the opposition encountered by Baily in 1811 in a matter which his own work and experience had shown him to be important, may have acted as a powerful stimulus when he met some of the same opponents later over *Nautical Almanac* matters.   Thomas Young was the Secretary to the Board of Longitude, which had allowed the *Nautical Almanac* to fall behind the times : and other influential names may have been common to the two controversies. Baily was a stalwart champion, as he showed later, especially in the case of Flamsteed ; and it is possible that our Society benefited to some extent by the warmth which his opponents had stirred within him in 1811, and which had not yet cooled in 1820 when it aided in the hatching of the Astronomical Society.

Baily was the backbone of the Society throughout its early years, as we shall be reminded in dealing with the time of his death in 1844 (see third decade), and in the éloge by Sir John Herschel pronounced on that sad occasion ; but we may add here a few details concerning the voyage to America (1795–7), which is indeed mentioned in the eloge, but scarcely so as to give a just notion of its character.   Indeed, even his intimate friends do not appear to have been aware of what was involved ; though Baily had kept a careful journal he did not apparently talk to them about his experiences.   " He was more than commonly reserved in matters relating to himself : and no old soldier was ever more chary of referring to anything which would insinuate dangers faced or hardships endured."   So wrote Augustus De Morgan in 1856, in editing the *Journal* which had been put into his hands. " In the course of fourteen years of intimate acquaintance I never arrived at so much knowledge of his adventures as is contained in the few sentences which formed the sum total of Sir John Herschel's recollections " (this is an allusion to the paragraph in the éloge of which mention is made above).   " Occasionally, when some thriving city was mentioned, he would say, ' When I passed that spot it was all forest,' or the like ; but I never heard him drop a hint that he had calculated, under those trees, the chances of being scalped or starved.   From all I knew of the writer, I feel sure that the hardship and the risk are both understated."

The risk of our losing even the understatement was also considerable.   De Morgan, after some consultation with Airy and others, put the journal through the press under the title, *Journal of a Tour in Unsettled Parts of North America in 1796 and 1797.*   By the late Francis Baily, F.R.S., P.R.A.S. (London, Baily Brothers, Royal Exchange Buildings, 1856), but he did not put a copy in the R.A.S. library, and we owe the copy we now possess to the kindness of his son, William (on the occasion of his dining with the R.A.S. Club, 1911 Nov. 10). It is rendered the more valuable by having inserted in it various maps found with the journal, some pamphlets by Baily, some letters, and Baily's attendance card for the Cambridge meeting of the British Association in 1833. The maps, so innocent of detail, fully bear out the modest remarks of the traveller just quoted. But a more concrete illustration of what travel meant in those days may be quoted from the book. On 1796 December 10, the boat in which Baily and his companions were going down the river Ohio got frozen in near the bank—a situation they accepted with tolerable equanimity. They prepared to pass the winter there and proceeded to lay in a good stock of provisions. One or two other boats were with them, so that there were fourteen or fifteen in all, and there was a daily expedition to shoot " deer, turkeys, bears, or any other animals fit for food." The temperature was about 17° below zero. On December 20 they thought there seemed some promise of the ice breaking and allowing them to proceed on their journey, and accordingly went to bed in fairly good spirits. Suddenly they were awakened by a noise like thunder, and found that the ice was indeed breaking up, but because the river was rising rapidly. We have seen that Baily was not a man to make too much of any experiences of his own, yet this is how he writes about the matter :—

> All attempts would be feeble to describe the horrid crushing and tremendous destruction which this event occasioned on the river. Only conceive a river near 1500 miles long, frozen to a prodigious depth (capable of bearing loaded wagons) from its source to its mouth, and this river by a sudden torrent of water breaking those bands by which it had been so long fettered ! Conceive this vast body of ice put in motion at the same instant, and carried along with an astonishing rapidity, grating with a most tremendous noise against the sides of the river, and bearing down everything which opposed its progress ! the tallest and the stoutest trees obliged to submit to its destructive fury, and hurried along with the general wreck ! In this scene of confusion, what was to be done ?

The practical answer was to unload the boat and get the things ashore as quickly as possible ; and this, in the middle of this bitter night, they set about ; but suddenly a large sheet of ice stove in one side of the boat so that she filled and sank, but—

> as she was near the shore and almost touched the bottom (the water being very low) she was not immediately covered. The river was

rising at a very rapid rate, and as we knew that if we once lost sight of her, we should never see her more, . . . jumping into the boat, up to our middle in water, we continued to work near three hours . . . the thermometer was at 17° below zero, and so intense was the cold that . . . the moment we raised our legs above the water (in walking) our stockings froze to them as tight as if bound with a garter ! In such a situation . . . it is a wonder we had not perished ; and possibly we might had not the river . . . completely covered our boat and obliged us to desist. Thus went our boat ! and thus went every hope of proceeding on our journey ! Thus were all our flattering prospects cut short, and none left but the miserable one of fixing our winter habitation on these inhospitable shores.

When day appeared they found that their troubles were by no means over. They had landed the gear on level ground, but above this the bank was " fifty feet high and nearly perpendicular," and the water rising rapidly. They carried the things up one by one, a few feet at a time, and lodged them behind some tree to prevent their rolling back into the river. When it had ceased rising they set about fixing up some kind of habitation out of poles and blankets, and kept a fire burning. " Some of the packages were so much frozen as to take three days standing constantly before the fire ere we could get out their contents to dry them." Presently they found a rough log hut which gave better shelter ; and they got a boat built for proceeding on their journey when the river (which froze up again, having been broken up merely by heavy rains) should allow, which was on February 20. The entry in the journal on Christmas Day is somewhat in the manner of a Robinson Crusoe soliloquy.

> Here am I in the wilds of America, away from the society of men, amidst the haunts of wild beasts and savages, just escaped from the perils of a wreck, in want not only of the comforts, but of the necessaries of life, housed in a hovel that in my own country would not be good enough for a pigsty ; at a time, too, when my father, my mother, my sisters, my friends and acquaintances, in fact, the whole nation, were feasting upon the best the country could afford.

Yet he comforts himself with the reflection that he was at least better off than those who had perished in the flood, or even than his companions who were at that moment ill with their experiences, though this meant that he was the only one able to work. And of course he survived these and other hardships as we know, and lived to nurture our young Society.

It appears that " one of the objects of his tour was the formation or extension of commercial connection, probably of some house in England.   It also appears that during his voyage he gave formal notice of his intention to apply for the privileges of citizenship, with a view to take up his permanent residence in the United States " (*M.N.*, **14,** 112), but fortunately for us nothing came of this.   On his return to England he seems for a year or two to have contemplated a life of travel and adventure, and we must admit that he had good qualifications which seem to have been shared by other members of his family.   Two of his nieces were the first ladies to make the journey up the Nile to the second cataract above Wady Halfa, and one of them, Miss Ann Louisa Baily,* survived to the age of 100 (on 1917 January 17).   Again, fortunately for us, these plans came to nothing, and about the end of 1799 he was taken into partnership by Mr. Whitmore of the Stock Exchange.   He meant so much to us that the temptation to linger with him is strong ; but reluctantly we must pass on, referring those interested to Sir John Herschel's éloge, with the supplement by De Morgan in *M.N.*, **14,** 112.

The beautiful bust which stands near the staircase of the R.A.S. rooms is the work of Edward Hodges Baily, and was presented to the Society by Miss Baily, the surviving sister of Francis Baily, on 1849 February 9 (*M.N.*, **9,** 53).

His younger brother, Arthur Baily, was also an original member of our Society, but is a rather shadowy figure.   He was born " about 1787 " (the birthday of Francis is known exactly as 1774 April 28), and died on 1858 July 8.   For two years he was Treasurer of the R.A.S. Club, but his enthusiasm seems to have been chiefly reflected from his brother.

It is scarcely necessary to say anything here of the Herschels or of Babbage and of the remaining original members.   Sir James South will appear later, since he became so conspicuous when the Charter was applied for.   But a word may be said of Peter Slawinski, a Pole who happened to be in England for some years, and is included in the original list as a Fellow, but may perhaps more appropriately be regarded as our first Foreign Associate.   He was appointed Director of the Vilna Observatory in 1825, but he took a part in our early meetings and presented a medal of Copernicus to the Society, which we still have (in the safe).   The following extract from the *Bulletin de l'Observatoire de Vilno*, 1921, gives a glimpse of him :—

* By the kindness first of Mr Hollis and then of Mr R. E. Day, the writer had the rare pleasure of an interview with this fine old lady on 1917 July 4 ; and to his delight was presented by her with a copy of Francis Baily's *Journal*, which had belonged to her sister, Emily (her companion on the Nile voyage). She was alert about the war. She " did not *want* us to kill the Kaiser ; but she would not much mind if his own people killed him."   She died at Esher on 1917 October 23.

Sniadecki remained Rector of the University until 1815 and Director of the Observatory until 1825. He increased the value of this Observatory by importing many new instruments.*

After *Sniadecki*, his disciple, *Pietr Slawinski*, took his chair and the direction of the Observatory. He made many observations. Among other things he determined the latitude of the Observatory ($= 54° 40' 59''\cdot 1$), and we find this value in contemporary astronomical almanacs. He also took share in geodetical measurements. In 1828 he published a handbook on theoretical and practical Astronomy. In 1832 the Russian Government closed the University, but the Observatory remained and was entrusted to the Academy of Sciences in St. Petersburg. . . . *Slawinski* retired in 1843. . . . In 1876 a fire broke out in the Observatory and caused some damage; then the Government closed the Observatory and gave its instruments and most of the valuable library to the Observatory of Pulkowa.

If we may venture a little outside the original list, it is worthy of remark that the Society seems to have owed a good deal in its early years to explorers and travellers, like Franklin and Parry. Franklin was away at the actual time of the foundation on his " Voyage of Discovery to the Northern Coast of America " (1819–22), but he seems to have joined almost immediately he returned, for he is entered in the Treasurer's Index as having compounded. Parry is entered as " non-resident," and, indeed, he was away on two voyages in 1819–20 and in 1821–23. We have the accounts of these in our library, but apparently not that of Franklin : yet his adventures in America and those of his companions remind us no less than those of Francis Baily how much has happened in the intervening century. We venture on one sentence from Franklin's book :—

> Hepburn having shot a partridge, which was brought to the house, the Doctor tore out the feathers, and having held it to the fire a few minutes, divided it into seven portions ; each piece was ravenously devoured by my companions, as it was the first morsel of flesh any of us had tasted for thirty-one days,

during which they had been feeding on lichen scratched off the rocks.

We may also note the instructions given to Captain Basil Hall and other navigators in the first year of the Society's existence. Another figure associated with these adventurous days was that of the Rev. George Fisher, who was appointed astronomer to the expeditions which set out for the Arctic in 1818, to make pendulum

---

* Sniadecki was in England in 1787, went to Slough and took lodgings there in order to see some object through Herschel's telescopes. " He was a very silent man " (*Memoir of Caroline Herschel*, p. 75).

determinations and do other scientific work; but the vessels *Dorothea* and *Trent* encountered a violent gale and had to put back. Nevertheless, Fisher swung his pendulum in Spitzbergen and got good results. In 1821 he was appointed astronomer (and chaplain also) to Parry's Expedition in search of a North-West Passage, and he prepared the scientific report on this expedition in 1825. In 1834 he took charge of the Greenwich Hospital School with great success, and became a constant attendant at the meetings of the Society and the dinners of the R.A.S. Club till his death in 1873.

A picturesque figure of a different kind is that of Dr. Lee of Hartwell, and an excuse for mentioning him here may be found in the fact that though he was only the second treasurer of the Society, and did not succeed Dr. Pearson until 1831, it is due to him that we have recovered a copy of the original accounts. On receiving them from Dr. Pearson he had them copied * very neatly, and apparently returned the original, which has in any case disappeared. But the copy remained among his books, and was ultimately presented to the Society a few years ago by Mrs. Lee, the relative into whose possession they had passed, on an occasion when it became necessary to remove the library from Hartwell. The volume published under the title *Speculum Hartwellianum* renders it unnecessary to dwell on the astronomical activities of Dr. Lee or of Admiral Smyth, who observed with him, but a sidelight is thrown on the old gentleman by the following extract from *Reminiscences* (Macmillan, 1922), by Constance Battersea (a daughter of the second son of N. M. Rothschild, who married (1877) Cyril Flower, created Lord Battersea in 1892) :—

> Hartwell House was an attraction to us in our young years, and we used periodically to visit its strange old owner, the learned Dr. Lee. He would take us into his wonderful and crowded museum, where on one occasion he presented me with a little stuffed bird, hoping that it might prove the forerunner of a collection of my own, for he said that the pleasure of collecting, no matter what, was one of the chief roads that led to a happy life.
>
> On one unforgettable starlight night we were admitted to his famous observatory and invited to look through the telescope, receiving much valuable information at the time. To us the astronomer seemed a very old man, and when, in company with one of his own years, he came to dine at Aston Clinton, we were not astonished at seeing the two aged guests of my parents dropping placidly off to sleep after their repast.
>
> The Hartwell estate has been in the possession of one family—

---

* Perhaps by his secretary, Mr. B. Smith ? See p. 21.

the Lees—for seven hundred years. Part of the house is of very old date, but I believe it was almost entirely rebuilt in the seventeenth century. One fact will always add greatly to its interest : from 1809 to 1814 Hartwell was the abode of Louis XVIII., the exiled King of France, and his queen, who died there ; with them came about 180 persons belonging to their household and Court. Many distinguished names, such as those of the Royal Dukes of Berry and of Angoulême, also those of Duras, de Gramont, de Servant, de Blacas, that of the Archbishop of Rheims, etc., became well known in the neighbourhood. Indeed, several of the monarch's companions died at Hartwell and are interred in the little churchyard belonging to the place.

I have read in Ditchfield's *Memories of Old Buckinghamshire* how " the halls, gallery, and larger apartments were often divided and subdivided into suites of rooms for the use of the members of the French Court and household, in some instances to the great disorder and confusion of the mansion. Every outhouse and each of the ornamental buildings in the park that could be rendered capable of decent shelter was densely occupied. It was curious to see how some of the occupants stowed themselves away in the attics of the house, converting one room into several by the adaptation of light partitions." Moreover, I have been told that a garden was laid out on the roof, where, besides shrubs and flowers, vegetables were grown for the use of the inmates.

PLACE OF MEETING

From the Minutes quoted we see that the preliminary meeting was held at Freemason's Tavern, and others followed in the hospitable rooms of the Geological Society. In 1820 November the Society moved into rooms in Lincoln's Inn Fields, paying an annual rent of fifty guineas to the Medical and Chirurgical Society, who allowed them, besides the room for meeting, the use of an attic. It was not, however, until 1828 that the Society availed itself of this luxury : in that year a difference with the printers caused them to remove their stock of publications from their care, and the attic became a storehouse. Meanwhile, many attempts had been made to find a more suitable and more permanent home. In 1823 a large room in the Scottish Hospital, Crane Court, Fleet Street, was reported, but on enquiry was found to have been already let. In 1824 March a possible house in Lincoln's Inn Fields was found to be priced at 6000 guineas, and judged, therefore, too large and expensive. In 1824 May the Council resolved—

> That the Secretaries do insert an advertizement in *The Times* and *Morning Chronicle* Newspapers for a House containing at least one room of about 30 feet by 20 feet in dimensions as a meeting-place for the Society ; such house to be situated between Tottenham

Court Road and Bedford Row, and not more than a quarter of a mile north of Holborn.

These details of the Society's needs a century ago have an almost comic aspect to us to-day: modest as they were they could not be met. Only two answers to the advertisement were forthcoming: one concerned a house in Newgate Street, which was put aside without even the formality of an inspection; the other was to be inspected, and presumably was found quite unsatisfactory, as there is no further mention of it. In December of the same year, Mr. Millington, the Secretary, was requested to confer with the managers of the London Mechanics' Institution in Southampton Buildings, but it was reported that the premises were quite unsuitable. A new Committee on premises was appointed in 1828 April, and soon afterwards (1829 November) the possibility of finding room in Somerset House began to take shape, ultimately becoming a reality in 1834.

THE MEMOIRS

No more important decisions were taken in the early years than those for the printing of papers read to the Society. Regulations for printing papers were adopted on 1821 May 11. An abstract of each paper, made by one of the Secretaries, was to be read to the Council, who should then decide by ballot whether the paper be referred to a Committee nominated by the Chairman: any alterations suggested were to be submitted by one of the Secretaries to the author (or his representative), and an estimate obtained for any plates required: and the Council were then to ballot for printing the paper as amended. We may note that " if such motion be carried in the affirmative, the author [was to] be allowed to correct the proof-sheets, if he should desire it." At a special meeting of the Council summoned for May 25, six of the papers already read were thus submitted to Committees, three of which reported favourably in June; and thus volume 1 of our *Memoirs* was prepared. It was resolved (in 1821 June) to print 500 copies of the *Memoirs*, with 25 extra copies of each paper supplied gratis to the author: members to be entitled to buy at half-price. The type was to be " that used in Mr. Colebrooke's *Indian Algebra*: and that the size of the page be the same. Which type is denominated by the printers, Pica: and the size of the page is $7\frac{1}{2}$ inches long and $5\frac{3}{10}$ inches wide." The advertisement which opens each volume of *Memoirs* was drawn up. The Minutes of this meeting conclude with a letter received from Sir Humphry Davy, President of the Royal Society, which may be transcribed :—

June 1, 1821.

SIR,—I beg you will return my thanks to the President and Council of the Astronomical Society of London, for the honour they have done me in sending me the account of their proceedings, in which every lover of science must take a strong interest.—I am, Sir, yr. obed. servt.,

H. DAVY.

To F. Baily, Esq., F.R.S., S.A.S.

We may take it, therefore, that the little friction of the year before was entirely at an end. The adopted regulations only differ in detail from those now in force, but one detail is important, viz. the preparation of an abstract by the Secretaries. This naturally caused delay, which was at times inconvenient : and in 1825 May revised regulations were adopted with a view of saving time. Firstly, the reading of the paper to the Society was taken as the occasion of the appointment of referees (instead of the reading of an abstract to Council) ; and secondly, if the referees and chairman were in favour of publishing the paper at once, they might proceed without waiting for a Council meeting, provided always that the matter be reported to Council at their next meeting. Otherwise the Council would ballot in the ordinary way. It is abundantly clear from the Minutes that this critical procedure was no empty formality. Thus on 1829 November 27, no less than five papers which had been read at the last meeting were rejected by the Committee and (after ballot) by the Council for printing. Any paper for the *Memoirs* is still subjected to ballot, and it is interesting to see how quickly our regular procedure was got into shape. It is true that most papers nowadays are printed in the *Monthly Notices* without ballot, but it is easy to see how a practice originally adopted for brief communications has been naturally extended, from its obvious advantages, to all but very long papers.

At the outset, however, the *Memoirs* were the only publications of any kind (with a possible modification mentioned below). The following notes may be found convenient :—

Memoirs, vol. i., is published by Baldwin, Cradock & Joy, and bears on the title-page the date 1822 ; but this title was issued with Part I. of the volume (pp. 1–232). Part II. ends with the 4th Report, dated 1824 Feb. 13, and the list of officers for 1824–5 ; and on the last page is the name of the printer (Richard Taylor, Shoe Lane).

Memoirs, vol. ii., same publisher and printer ; 1826 on title-page ; ends with Baily's Presidential addresses on 1826 April 14, and a list of Fellows on 1826, June 19.

But there follows as an Appendix, Baily's Catalogue of 3000 stars.

Memoirs, vol. iii., published by Priestley & Weale, Holborn ; 1829 on title-page.   Printed by Richard Taylor, now removed to Red Lion Court, Fleet Street.   Ends with list of Fellows on 1829 February 13.   (Minute of Council on this day states that the volume was closed there because of change of printer.)

Memoirs, vol. iv. ; Priestley & Weale, 1831 ; printed by J. Moyes, Castle Street.   Ends with list of Fellows on 1831 June 10.

By this time the *Monthly Notices* had been established, and we may now turn to their early history, showing incidentally how these changes of printer and publisher came about.

THE " MONTHLY NOTICES "

The first number of the *Monthly Notices* is headed with the date 1827 February 9, and contains the Council Report.   It has been generally accepted as indicating that there are no similar records for the first seven years of the Society's existence.

But in the copy of this number possessed by the writer there is the imprint " From the Philosophical Magazine and Annals ; printed by Richard Taylor, Shoe-Lane, London," which naturally led to an examination of the *Philosophical Magazine* of that date. It was found that brief notices of this kind had been regularly printed therein from the first : the novelty was simply that of having separate copies struck off for the use of the Fellows.   Even then there was apparently no thought of collecting them in volume form : this was not done till 1831, and most copies of the first fourteen *Notices* are reprints made about that date by a different printer (J. Moyes, Castle Street, Leicester Square, with Priestley & Weale as publishers).   The copy above mentioned, which apparently contains the sheets struck off at the time, exhibits one or two features of special interest.   Thus following No. 1 there are two copies of No. 2, both dated 1827 March 9, and identical except for two features : firstly, Roman figures (No. II.) are used in the first (as in No. I.) and Arabic (No. 2) in the second, and consistently afterwards : secondly, a whole sentence is omitted from the later copy.   After the words in No. 2 :—

> M. Gambart exhibits a comparison of the results of these elements, and of his observations on the 27th, 28th, and 29th of December,

there follow in the originals (*i.e.* in the *Philosophical Magazine* itself and in the first reprint) the words :—

> He then adds a few remarks, which need not be recorded, and congratulates himself and astronomers generally upon the existence and success of the Astronomical Society of London.   " What," he asks, " may not be expected from so liberal an association ?

Happy the country where the love of science alone causes so many men of enlightened minds to combine in such an object ! Happy, also, those who dwell there ! "

Apparently the modesty of some members of the Society was shocked by this public commendation and a new reprint was ordered with the requisite omission. But the words still stand in the *Philosophical Magazine*, and there is no harm in recalling the incident at this date.

The *Philosophical Magazine* was founded by Alexander Tilloch, who was joined by Richard Taylor as editor in volume lx. (1822). References to the Astronomical Society begin in volume lv. (1820) with the announcement of its formation (p. 147), the Address (p. 201), and a few words about the first ordinary meeting on March 10 (p. 225). Each monthly number of the *Magazine* consisted at that time of exactly eighty pages, and it is easy to find the references near the end of the number among the doings of other Societies. But if we turn over the leaves in years preceding the formation of the Society in 1820, we find that matters of astronomical interest were frequently included. Thus in volume l. (1817) there is a list of errata in the *Nautical Almanac* and a somewhat vigorous criticism of the compilers of that work, so that the Board of Longitude was a deaf adder before 1820. On p. 407 there is a paper by Count Laplace on the " Rings of Saturn," suggesting that the stationary appearance of parts of the ring (which we know nevertheless to be revolving) may be due to slight relative inclinations of separate rings to each other. In volume lii. we note that a useful monthly list of astronomical phenomena given in previous volumes was for some reason dropped, though the meteorological information which regularly followed the list is retained. In volume liii., however, we get information about comets and ephemerides of the four minor planets. Such notes and references continued for many years, so that a student of astronomical history about the beginning of the century should not neglect the *Philosophical Magazine*, or, as it became in 1827, the *Magazine and Annals of Philosophy*.

Tilloch was sole editor in 1820, and the early *Notices* are thus due to him. But on the arrival of Richard Taylor in 1822 as joint editor, they become distinctly fuller, in which we may probably trace his influence. But it was not until 1827 that separate copies were struck off for distribution, and apparently the move came from within the Society. The Council Report of 1828 February 8 (*M.N.*, **1,** 49) seems clear on this point :—

One of the first acts of the Council of the year elapsed, was to enter into an arrangement with Mr. Taylor, the printer to the

Society, and who is also one of the editors of the *Philosophical Magazine*, for the publication of a series of monthly notices of its proceedings, and for the supply of a sufficient number of copies of them, in succession, for distribution among the members. The convenience and advantages of this plan have been sufficiently proved by the trial which has been given it, and it will, of course, be continued.

[There follows a summary of the advantages of the plan, which need scarcely be reproduced here, as the passage is readily accessible.]

But before the first volume was completed differences of opinion arose between the Council and the printer on the question of cost of printing, and also about insurance of the stock, which, as we have seen, they took possession of themselves and housed in their attic. Estimates were obtained from John Weale, trading as a publisher under the style of Priestley & Weale, and were accepted. Moyes was the printer employed by this firm, and a letter from him is entered on the Council Minutes stating that he fully understood that he was to have no separate claim on the Society. Weale's estimates are given as replies to an elaborate series of questions. Part I. of volume **3** of the *Memoirs* was already set in type (by Taylor), and was sent with enquiries what would be the cost per sheet (for 500 copies)—

if all were like p. 132 . .	. Reply £1,	16s.
if part were tabular, like p. 38	. ,,	2, 10s.
if all were tabular, like p. 70 .	. ,,	3, 4s.
,, ,, p. 35 .	. ,,	5, 2s.

and so forth : and then it is asked what would be the charge " for 200 copies of the *Monthly Notices*, including paper, it being understood that you may dispose of the copyright as you may think proper " ? Answer, " This I would undertake to give the Astronomical Society as a return for their copy and copyright."

Weale's estimates were accordingly accepted, and the arrangement with the former printer (Taylor) and publisher (Baldwin) terminated in 1828–9. The dates for different transactions naturally vary a little. Taylor finished printing volume iii. of the *Memoirs*, but it was handed over to Weale for publication. The last *Monthly Notice* printed by Taylor is that of 1828 June, which is so lengthy that it was divided into two for printing in the *Philosophical Magazine*. The November *M.N.*, printed by Moyes, reappears indeed in the *Philosophical Magazine*, but not till 1829 March : that for 1828 December not at all. A few extracts are given from the January *M.N.* in the *Philosophical Magazine* for 1829 June, and nearly all the February *M.N.* (Council Report) in *Philosophical Magazine* for 1829 July. But the reproduction becomes irregular from this time.

We may now give an extract from the Council Report of 1834 February 14 (*M.N.*, **3**, 21), which shows how the Society finally took this useful publication under its own wing :—

> The second volume of the *Monthly Notices* of the proceedings of this Society has been terminated rather more suddenly than was anticipated, in consequence of the disinclination of the publisher to continue that work on terms originally proposed. The Council, however, finding that many fellows were desirous that the *Notices* should still be carried on, have directed them to be printed in future at the expense of the Society ; which, although an additional annual charge, will, they trust, be approved by the meeting. The several numbers, as they appear, will be forwarded as usual to Fellows residing within the limits of the threepenny post.

The *Monthly Notices* thus had a somewhat chequered origin. It is a little sad that they should have so early been separated from the printer who started them ; but it is some satisfaction to know that volume **1**, though it bears the name of another printer and publisher, begins as a transcription from the *Philosophical Magazine*, and can be underpinned by similar reprints back to the foundation of the Society. These have been copied out, and may perhaps, someday, be published as a volume zero.

It is of interest to note a further few details in connection with the early *Notices*. The type originally used was larger than that used subsequently, and the fashion of printing at the foot of the page the first word of the next page was retained in the original issue of the first three *Notices* and then dropped. It was never used in Moyes' reprint. The *Philosophical Magazine* uses the spelling phænomena, which is reprinted as phenomena ; and the capitals of Continental Astronomers disappear in the reprint. Such changes were apparently made by the printer and not by any reviser : for there is a sentence at the opening of the number for 1828 June 13, which should have caught the eye of any careful reader, viz.—

> From a mean of 15 measurements, he makes the apparent distance on the left side equal to $11''\cdot272$, and on the right side equal to $11''\cdot390$; the difference is $0''\cdot215$.

There is clearly a slip here, but the occasion of the reprint for revising it was not taken. The figures $11''\cdot272$ and $11''\cdot390$ are the results for the first day, copied inadvertently, instead of the means, $11''\cdot073$ and $11''\cdot288$.

By a curious coincidence the origin of the *Monthly Notices*, like that of the Society itself, was embarrassed by a Royal death. Volume **1** opens with the February meeting 1827, and the *Philosophical Magazine* contains no reference to a January meeting. The minutes of Council show that it was not held in consequence of the death of the Duke of York. The Council met at four o'clock, and the President (Baily) "announced that in consequence of the death of the Duke of York, he had deemed it necessary to put off the ordinary meeting, which was to have taken place this evening (1827 January 12), and had in consequence given the earliest notice thereof to the members through the medium of the public newspapers, the 'Times,' 'Post,' 'Chronicle,' and 'Courier,' and further stated that he had not postponed the meeting of the Council, it being absolutely necessary that Auditors should be appointed without delay, and the Report prepared for the Annual General Meeting in February."

Colby and Sheepshanks were accordingly appointed auditors, and the President and Secretaries requested to prepare the Report. In these days the appointment of auditors is no longer a matter for Council, but for the Fellows generally: and the Report is prepared without any special request.

THE MEDALS

In 1826 February a somewhat distressing situation was reported to the Council. A number of medals had been struck (at the Mint, as was decided in 1823 December) from the dies for future use, and should be in their possession, but could not be traced, the deficiency being no less than three gold and seventeen silver medals. Mr. Millington (who had been Secretary at the time of the receipt of the medals, but had since resigned) had been written to more than once, but no reply had been received from him. However, the alarm was needless. By the next meeting of Council the dilatory correspondent had replied that he had indeed in his possession the "silver proof" medal which had acted as pattern, but that the others would probably be found "in the place in which they were deposited": and although this was not further specified in the Minutes, possibly for prudential reasons, the missing medals were produced and laid on the Council table. The Council in their relief did not forget to direct formal application to be made to Mr. Millington for the proof medal, which was ultimately also recovered.

Comparison of one medal with another revealed some difference in weight: thus two gold medals were found to have a mean weight of 2 oz. 2 dwts. 12⅔ gr., whereas one previously awarded to Mr. Babbage had weighed 2 oz. 7 dwts. 20 gr., the difference of

5 dwts. $7\frac{1}{3}$ gr. in weight, corresponding to £1, 7s. 7d. in value. It was resolved that the medals in future should have a uniform value of £10 of the currency, and about the same time (1827 November) it was resolved that all proposals for medals should be made in December and considered in January, an approximation to the procedure which ultimately took shape.

In the earlier years proposals for medals had not followed any regular rule, and had been more numerous than they ultimately became. It was one of the intentions of the newly formed Society to stimulate the progress of Astronomy by proposing prize questions for solution, but no great success attended these efforts. The questions proposed were often fully worthy of attention, or so we should judge in the light of our riper experience. Thus in 1824 April it was decided to offer a silver medal for a stellar photometer, and in response a suggestion was made by one of the Greenwich assistants, but apparently not approved. Many years were to elapse before a satisfactory solution of this problem was approached; but looking back now, from our presumably deceptive standpoint, it is puzzling to think that the need was formulated and attempts made to meet it, with so little success. With regard to other problems suggested at the same time, such as the provision of tables of the newly discovered minor planets, and the integration of the equations in the problem of three bodies, we can understand the reasons for failure in response; but why should not someone have devised some kind of photometer, even in 1824?

The first medals were presented at the Annual General Meeting of the Society on 1824 February 13, though the awards had already been announced at the previous November meeting. Two gold medals were given, one to Babbage for his (first) calculating machine, and one to Encke for his determination of the elliptic orbit of the comet called after him. Two silver medals were given to Charles Rümker and to Pons, both for discoveries of comets. We may here refer to the list of recipients of the medals given at the end of this volume, or to the lists of the awards printed at the end of nearly every volume of the *Memoirs* down to the present time. We shall here merely mention that silver medals were only given to two other recipients, to Stratford and to Beaufoy, both in 1827. In 1826 and 1827 the medals were presented at Special General Meetings in the month of April; but beginning with 1828 they have always been presented at the Annual General Meeting in February, though this is merely a custom and is not a consequence of any bye-law. The presidential addresses delivered on these occasions, which at first were very short, became gradually longer, and have always been valuable and interesting summaries of the work done by the medallists.

THE SECRETARIES

If the Council worked hard at the outset the Secretaries must have been taxed severely, and generally found their labours too serious after a few years.   Even the energetic F. Baily, who (with Babbage) initiated the office, only held out a couple of years.   He usually signed the Minutes as Secretary (in addition to the signature of the Chairman); but there are several omissions, and the Minutes of 1820 December 8 were signed on 1821 January 12 by C. Babbage as Secretary (though Baily was present on both occasions).   He gave notice in 1822 March of his desire to resign at the end of the session (June), and at a special meeting held on November 1, Millington was elected Secretary, and Baily invited to continue attending the meetings of Council " that they may avail themselves of the benefit of his advice " : he accordingly did attend, the fact being specially mentioned on each occasion. The handwriting (uniform up to that point) then changes, and the minutes are not afterwards signed by either Secretary.

Babbage also wished to resign in 1822 February, but agreed under pressure to accept office with Millington.   The early holders of the office were :—

F. Baily, 1820–23.	O. Gregory, 1824–28.
Babbage, 1820–24.	Stratford, 1826–31.
Millington, 1823–25.	Sheepshanks, 1828–31.

In 1824 March (at one of the resumed Council meetings after the evening meeting) it was resolved " that in consequence of the increased business of the Society in correcting press and various other ways, it has become necessary that an Assistant Secretary or clerk should be employed, and Mr. W. S. Stratford, R.N., being recommended by Mr. Gompertz and Mr. Frend as a person highly qualified to fill this office, was appointed Assistant Secretary to the Society from this day until the commencement of the vacation in June next, and that he be remunerated for his services in such manner as the Council shall determine."

Knowing the present importance of the office thus initiated, we learn with surprise that the start was in this instance not followed up.   In 1825 May, Millington resigned one of the Secretaryships and Stratford (who was paid in all £40 for his services) offered to fill the gap in an honorary capacity.   The offer was accepted, but Stratford's name does not appear in the lists of those present at the meetings until 1826 March, when he had been regularly elected Secretary.   In 1825 November, however, he was elected a Fellow of the Society, and in consideration of the " close and unremitting attention which he had constantly paid to the duties of the office of Honorary Secretary," the Treasurer was directed to regard him as a Life Member, without payment of fees. There is no further mention of an Assistant Secretary until 1829 January 29, when we find it—

Resolved that the Secretaries be empowered to employ an Assistant.

But nothing more definite occurs in the Minutes until 1830 November 19, when steps were taken resulting in the appointment of Mr. James Epps, as recounted in the next Chapter.

STANDING COMMITTEES (INSTRUMENTS AND LIBRARY)

In 1829 January two Committees were appointed to report upon the Instruments and on the Library respectively. They were to be reappointed in each following January. The latter gradually became a permanent institution, but there is to-day nothing in place of the former. It may therefore be well to recall the importance of this matter of instruments in the early days. The first considerable present of instruments was from Lieut. George Beaufoy. who at the death of his father, Colonel Mark Beaufoy (a silver medallist of the Society in 1827 February for his observations of Jupiter's Satellites " with a five-feet achromatic by Dollond ") handed over

> One 4-feet transit by Cary,
> One altazimuth by Cary,
> A sidereal and a mean solar clock,

in recognition of which valuable presents the donor was elected a Life Member, without payment of fees (*M.N.*, **1**, 51). At the adjourned evening meeting of Council after the presentation, 1827 June 8, Captain Smyth, R.N., applied for the loan of the instruments. His request was referred to the Committee of the President (J. F. W. Herschel), Beaufort, Baily, and Colby, which had already been appointed to deal with the handing over of the instruments and is referred to as the " Instrument Committee." This no doubt led to the establishment of the more permanent Committee some eighteen months later. In 1828 December 8, Dr. W. H. Wollaston presented a telescope made by Peter Dollond in 1771 : expressing the hope that it might be used. He had himself used it for " trying and perfecting his method of adjusting the triple achromatic object glass " : and it was forthwith lent to Mr. Maclear, of Biggleswade, for observing occultations. Dr. Wollaston was proposed as a Fellow in 1828 June, and would in the ordinary course have been balloted for in December ; but the " alarming state of his health and high probability of his dissolution previous to the December meeting induced the Council to recommend to the meeting a departure from the established rule," and he was unanimously elected on November 14, making his present on December 8, and dying on December 22 (*M.N.*, **1**, 102–4).

Another valuable present of a 2-feet altitude circle divided on gold, which had belonged to the Rev. Lewis Evans, was made by Dr. Lee. Thus there was work for an instrument Committee, and they drew up regulations, approved by the Council, for the use of the instruments (*M.N.*, **1**, 121). In these printed regulations provision is made for marks of identification, but a special addition was made by Council on 1829 February 6, which may seem to us to-day a little drastic, viz. :—

> That on both surfaces of an object-glass of a telescope there be written with a diamond the words " Astronomical Society of London."

The addition is represented by the phrase in brackets in regulation 2, *loc. cit.* ; and Dr. Wollaston's telescope was at once sent to Mr. Dollond to be marked accordingly.

In connection with the Library Committee we may note De Morgan's offer to the Council in 1829 June, to " form a Catalogue of the Society's Books," which was gratefully accepted.

THE BOOK OF SIGNATURES

In 1828, William Henry, Duke of Clarence, Lord High Admiral, was elected as the 301st Fellow of the Society, and the occasion seems to have been taken to purchase the book in which the Fellows sign their names at the admission ceremony, for we have a note of Booth's charges for an " autograph book."

The name of the Duke of Clarence was the first signed in this book, but it stands as William IV. The entries on the Royal page are :—

William R., Patron.	Victoria, R.I.
William.	George, R.I.
Augustus Frederick.	

These names are subscribed immediately under the following declaration, from which they might have been supposed specially exempted :—

> We, the Undersigned, being elected Members of the Astronomical Society of London, do hereby promise that We will be governed by the Regulations of the said Society as they are now formed, or as they may be hereafter altered, amended, or enlarged. That we

will advance the objects of the said Society as far as shall be in our power : and that We will attend the usual meetings of the Society as often as we conveniently can.   Provided that whenever We or any of us shall signify in writing to the Society that We are desirous of withdrawing our names therefrom, We shall, after the payment of any annual contribution which may be due by us at that period, be free from this obligation.

After a few blank pages the three following names and date are entered on a separate page for some unknown reason :—

> El Conde de Montemolin.
> June 11, 1847.
> General Montenegro.
> Le Chevalier de Berard.

No mention is made of these gentlemen in the *Monthly Notice* for 1847 June 11, nor in the index to the volume of *M.N.* : the only Associate elected recently was Le Verrier.   It may be presumed that three distinguished visitors were present at the meeting and were asked to inscribe their names on a page which lay just outside the list of Fellows.   This begins on the following page, room being left at the top of the page for the declaration to be filled in (on each page) ; but this has never been done.

The first names in the list of Fellows are those of H. T. Colebrooke, Francis Baily, John F. W. Herschel, Davies Gilbert, William Pearson, etc. etc.

After the acquisition of the Royal Charter a new declaration was written out and a new list of signatures was started on p. 7, many Fellows signing afresh (*e.g.* F. Baily, B. Gompertz, and Edward Riddle occur in the first dozen names of both lists).

The first name of the new list is that of Francis Baily.   But the existence of two lists has produced some confusion, the book having apparently been opened on more than one occasion at the wrong place ; for the name preceding Baily's (*i.e.* the last on p. 6) is the modern one of

> E. B. Knobel for Mr. J. H. Honeyburne.

On p. 8 someone at some time has upset the ink, but all the signatures show through quite well.

ASSOCIATES

It has already been remarked that we may regard Peter Slawinski as the first Associate Member.   The first Associate to be elected in the ordinary way was Biot, proposed in 1820 June and elected in December.   Seven others were proposed in November and elected in January ; and the list of Associates printed on 1821 February 9 thus contains nine names :

Jean Bapt Biot, Paris.	William Olbers, Bremen.
Alexis Bouvard, Paris.	Peter Slawinski, Wilna.
John B. J. Delambre, Paris.	J. D. Vallot, Dijon.
Chas. F. Gauss, Göttingen.	Hen John Walbeck, Åbo.
C. Louis Harding, Göttingen.	

It is possible that the presence of Slawinski in England at the time of the original meeting, and his presence among the fourteen who met on January 12, may have suggested the status of Associate Members ; but it will be noted that no distinction is drawn between his name and the others in the Minutes of January 12 and those of February 8 ; the numeration continues in this sense.

APPENDIX TO DECADE 1820–30 (Chapter I)

(1) Although the project of this History has been kept in mind for several years in order to make as complete a research as possible, it is inevitable that some references should only be discovered just too late.   After the MS. had been sent to the printer Miss Herschel kindly sent me a scrap of a letter from Sheepshanks to Sir John Herschel.   Judging by another scrap, which implores Sir John to burn the letter (of which accordingly little more survives beyond this injunction), the remainder of the letter below was probably destroyed, including the date ; but it was almost certainly written early in 1848, when Sheepshanks must have been writing the obituary notice of Pearson, printed in *M.N.* **8**, 69.   The letter and the notice shed light on one another, and the letter is valuable as emphasising the difficulty that was found, less than thirty years after the foundation, in recovering the exact details of our early history.   Pearson and Baily were both dead, and they alone appear to have known the facts.   Sheepshanks apparently took great pains to ascertain them, and might have hoped to get information from Sir John Herschel, if from anyone, but apparently the attempt failed.

The following is the portion of the letter referred to :—

. . . obedient of slaves, it is most conspicuously seen when I am ordered to do what I like.   Seriously I think all these proposals *are good so far as they go*.   I should object exceedingly to stepping out of our proper business, but I see no harm in a modest suggestion which binds the advised person to nothing, and which is so indirect that it need not be heeded except by a willing person.   I am not sure whether the reticentia of good and sensible men is not the cause of much of the mischief done by charlatans.   We blame people for being humbugged, without considering that humbug has been the only pabulum offered to them. So please let us to some extent consider ourselves a power, a very modest one, confining ourselves merely to discriminating praise, but assuming tacitly that we can discriminate and that our praise is worth having.

Thank you for your note with Babbage's recollections. I have tried to do justice to both Pearson and Baily, but I own that I have leant a little *against* my inclinations. I wish I could have made out who called the first meeting on Jan. 12, 1820. Stokes thinks with Babbage that Baily had the principal share, and remembers that he proposed Danl. Moore as one of the Committee of eight to keep out Dr. Kelly. If so much had been discussed in Blackman Street *before* this meeting as B. supposes, I cannot account for not including South in the committee of eight out of a meeting of 14. South was not in the first Council. Hence I must suppose that the active discussions at Blackman Street are later than the foundation. Perhaps we shall get more light in time. Best regards to Lady H.

Yours very dutifully,

R. Sheepshanks.

(2) After this chapter had been printed, Dr. Dreyer drew my attention to another instance of the ill-luck which befel Waterston (see pp. 27–28). In 1850 he published a paper " On a Graphical Mode of Computing the Excentric Anomaly " (*M.N.*, **10**, 169). About fourteen years later the method was re-discovered by Dubois (*A.N.*, 1404) and has been called by his name. The injustice was first noticed by T. J. J. See (*M.N.*, **56**, 54).

(3) To Dr. Dreyer I also owe a reference to Sprat's *History of the Royal Society*, from which it appears that a complete survey of the sky was not contemplated for the first time in 1820 (see p. 6), for " they (the R.S.) have suggested the making a perfect survey, map, and table of all the fixed stars within the zodiac, both visible to the naked eye, and discoverable by a 6-foot telescope with a large aperture. . . . This has been approved and begun, several of the Fellows having had their portions of the heavens allotted to them."

The date of Sprat's *History* is 1667.

CHAPTER II

THE DECADE 1830–1840. (BY J. L. E. DREYER)

1. IN 1830 February, when the Society had existed for ten years, it was generally felt that it ought to obtain a Charter of incorporation in order to secure the property already acquired or which might be acquired, and to insure the appropriation of that property to the particular uses for which it was destined.* A small committee appointed by the Council to look into this matter, reported the following month that they had learned that the expenses would to a great extent depend on the number of names introduced in the Charter. They had seen some recent charters, in which every necessary object had been fully attained by the insertion of one name only, that of the President, and of very few clauses. The amount of fees would come to about £270 and the law charges would not exceed £50.

Subscriptions to cover these expenses were therefore invited from the Fellows, and in a few months nearly £400 were collected. Eventually the expenses only amounted to £268, as Mr. Henry Hoppe transacted the legal business gratuitously.† In 1830 May the President (South, since 1829 February) was empowered to sign the petition to the King for the grant of a Charter, while a committee was appointed to examine the Bye-laws of the Society and to propose any alterations and additions that might become necessary. The matter dragged on ; in the following November it was reported to the Council that there had been some delay from unforeseen causes. But on January 14, South announced that he had attended the Levee on December 15, when the King inserted his name in the autograph book as Patron of the Society, and that in consequence the Society would now become the Royal Astronomical Society. At the same meeting of the Council the Charter Committee reported that they had settled the draft, and that a clause had been inserted, providing that no Fellow who had filled the office of President for two successive years

---

* This question had already been raised in 1825, but no effective action was taken.

† An unappropriated balance of £129, 18s. 10d. was handed to the Treasurer for the general purposes of the Society in 1832 February.

should be again eligible until the expiration of one year from the termination of his office.

This provision and the long delay in getting the Charter produced an awkward situation. To save expense, South's name only had been mentioned in the Charter, and in a very prominent manner. It begins thus :—

> Whereas Sir James South, of the Observatory, Kensington, in the county of Middlesex, Knight, has by his Petition, humbly represented unto us, that he, together with others of our loyal subjects, did, in the year 1820, form themselves into a Society. . . .

And further on it is ordered :—

> . . . . that the first members of the Council shall be elected within six calendar months after the date of this our Charter ; and that the said Sir James South shall be the first President of the said body politic and corporate, and shall continue such until the election as aforesaid.

Before the Charter was ready, the time of the Annual General Meeting of the Society came round again (1831 February 11). The draft of the Charter was read and approved, and Officers and Council were elected as usual. Although the unborn Charter said that South was to be the first President of the newly incorporated Society, it also ordered that nobody should be President for three years in succession ; and Brinkley, Bishop of Cloyne, was accordingly elected President.* But it must have been felt soon after, that this election of a President and Council was of doubtful legality. The Charter was at last signed by the King on March 7.† On the 19th the Council agreed to issue a circular, explaining that as doubts had arisen as to an informality in summoning a Special General Meeting for March 11, no business had been transacted on that day ; but that another General Meeting would be held on April 6 to decide on the acceptance of the Charter, and in case of such acceptance to elect Council and determine on bye-laws.

It was evidently a severe blow to South, that he was not to be the first President of the Royal Astronomical Society. He was present at two Council meetings in March, when Baily (in the absence of Brinkley) was in the Chair ; but he absented himself on April 6, when he got Stratford to announce that " he was desirous of retiring for the present year." As Barlow also wished

* Brinkley had vacated the Professorship of Astronomy at Dublin in 1827 on being appointed Bishop of Cloyne. It was said in Ireland that "he might thank his stars" for his promotion.

† Printed at the beginning of volume 5 of the *Memoirs*; also separately in 8vo in 1831 and several times, last in 1908.

to withdraw on account of being unable to attend, the names of Tiarks and Sheepshanks were substituted. At the Special General Meeting the same day the Charter was read, after which Baily quitted the Chair. "Sir James South, as the President named in the Charter, was then called for to take the Chair, and on its being ascertained that he was not present, it was resolved that Mr. Bryan Donkin be requested to preside," which he did. The Bye-laws were then read and submitted in sections for approval, after which the new Council was re-elected with the two modifications mentioned.*

South never again served on the Council, and after some years never appeared at the meetings of the Society. At the General Meeting in 1834, when no medal was awarded, he moved that the Council be recommended to give a gold medal to Stratford for the reconstructed *Nautical Almanac*; but an amendment was passed, censuring the irregularity of this motion. This was probably the last appearance of South at any meeting of a Society with many of whose members he was now at open feud. The cause of this was his quarrel with the firm of Troughton & Simms and the law-suit arising out of it. As this created a great deal of sensation and has been much misrepresented by South, it seems desirable to give a short account of it here.†

South after doing good work with smaller instruments, chiefly on double stars, erected an Observatory at Camden Hill, Kensington, about 1826. In 1829 he secured what was then the largest object-glass in existence, made by Cauchoix, of $11\frac{3}{4}$ inches aperture and 19 feet focal length. He entrusted Troughton with the task of constructing a mounting, and had a large dome built for it.‡ In the autumn of 1831, when the mounting was nearly finished,

* South had shown his ill-humour already at the February meeting, when presenting the medal to Kater, whose invention of the vertical collimator he "damned with faint praise." In the *Memoir of Aug. De Morgan*, p. 43, is a quotation from a letter (undated) from Smyth to De Morgan, in which he hopes "that the Council will take effectual steps to repel every disorderly attempt to impute motives or impugn its conduct"; but no particulars are given.

† In addition to the *Memoir of De Morgan*, the sources of the following account are, first, a pamphlet entitled *A Letter to the Board of Visitors of the Greenwich R. Observatory in Reply to the Calumnies of Mr Babbage*, by the Rev. R. Sheepshanks: London, 1854; 37 pp., 8vo. Secondly, the privately printed correspondence between South and Troughton & Simms, from 1832 May 16 to 1833 June 26, 84 pp. 8vo, interleaved and with thirty blank pages at the end; bound in boards. There is no title-page, but on the first page is the word "Appendix" in big letters, which word also forms the heading on every page. At the beginning South has written, "Most confidential."

‡ The telescope was erected on a temporary stand at the end of January 1830, and on February 13, John Herschel discovered with it the sixth star of the trapezium in the nebula of Orion. On the morning of May 14, Comet I. 1830 and Uranus, Jupiter, Mars, and Saturn were viewed with the "20-feet achromatic" (see *Monthly Notices*, **1**, pp. 153 and 180–181).

invitations were issued to prominent men to be present at the inauguration of the great telescope on November 26, by the Duke of Wellington. But the invitations were countermanded, as the shutters of the dome were unsatisfactory ; and though the telescope was erected in 1832 January, we do not hear of any ceremony having taken place. This was no doubt because South at once declared his strong dissatisfaction with everything. In a letter to Schumacher (dated June 22) he pronounced the dome to be " a national disgrace " and the polar axis " a decided failure." In the following July, South went to Dorpat to see for himself if Struve's equatoreal really was as steady as alleged, returning in the beginning of November. An acrimonious correspondence between South and Troughton & Simms, which had commenced in the previous spring, was now resumed with great vigour and, thanks to South's ill-temper, went from bad to worse. Every obstacle was put in the way of the experiments and attempts to strengthen the mounting, by which its makers endeavoured to rectify the faults found with it, a task which turned out to be hopeless owing to the utter impossibility of getting South to listen to reason.

The mounting was of the so-called English form. Judging by two sketches of it long afterwards published by Sheepshanks,[*] the upper and lower pivots were joined by what looks like two pair of semi-elliptical hoops. Each pair were in the middle connected by a St. Andrew's cross, in the centre of which were the supports of the pivots of the telescope, which was suspended like a huge transit instrument. The fault found with it was this : when the instrument was turned a little on its axis and then let go, a series of about a dozen short, quick vibrations followed, each lasting about 0·3 or 0·4 second. This was remedied by Sheepshanks by altering the bearing of the lower pivot. But now, when the instrument was turned in R.A. by laying hold of the telescope (it was quite steady when moved by the hour-circle) a vibration not unlike the former was discovered. This was obviously due to a twist in the great polar frame, in the construction of which no provision whatever had been made against twist. Sheepshanks therefore joined the two stays on each side of the telescope by pieces of wood parallel to the hour-circle ; bound these together with diagonal bracing, and finally covered the outside with a thin sheeting of boards parallel to the polar axis. " The instrument now looked as if it was made up of two decked boats, and in my opinion was handsomer than it was originally." Experiments with weights applied at the sides of the upper and lower ends of the polar axis showed that the twisting had quite disappeared.

* In the pamphlet issued in 1854.

The instrument had thus been freed from the imperfections complained of, but South remained obstinate, stopped further work and continued to refuse to pay Troughton's bill. Proceedings were therefore commenced towards the end of 1833 to compel him to do so ; but as so many technical questions were involved, the court recommended arbitration, which was agreed to. Mr. William Henry Maule, who was made arbitrator, had been Senior Wrangler in 1810 ; he became afterwards a Justice of the Common Pleas.* Counsel for Troughton & Simms was Mr. Starkie (Senior Wrangler in 1803), with Sheepshanks as his adviser ; for South was Mr. Drinkwater Bethune, a high wrangler of 1823 (Airy's year).† Maule at once insisted that Troughton & Simms should be allowed to finish their work according to the plan proposed by Sheepshanks, but only to be paid for if successful. In 1834 July, after most of the time allowed had been spent on getting a screw made and the clock put up,‡ the instrument was shown to and tested by Pond and Donkin, who had to acknowledge that it was perfectly fit for the work it was intended for, viz. micrometric measures of double stars. Measures of several pairs were successfully taken by Airy and others.

The legal proceedings went on for a couple of months longer, and in 1834 December the whole claim was awarded, including payment for the additions. But although the instrument had been proved to be satisfactory, South was not to be turned from his desire of posing as a martyr. He smashed the whole mounting to pieces, and in 1836 December advertised the fragments for sale by auction, by means of a scurrilous poster, in which the R.A.S., Troughton & Simms, and " their Assistants, Mr. Airy and the Rev. R. Sheepshanks," came in for a good deal of abuse. His folly cost him fully £8000. Lord Rosse offered to design and even to make an equatoreal for him, but he declined.§ We get an insight into his mind through a conversation he had with the American astronomer, O. M. Mitchel, in 1842. Exhibiting what he called the wreck of all his hopes (fragments of the mounting), South said in reply to a remark that the telescope might yet be mounted : " No, Struve has reaped the golden harvest among the double stars and there is little now for me to hope or expect." ‖

* It is a great pity that no record is left of the proceedings, as Maule is said to have been probably the greatest wit on the English Bench. For a delightful account of him, see *What the Judge Thought*, by E. A. Parry (London, 1922).

† Held afterwards a high Government appointment in India ; died 1850. Wrote lives of Galileo and Kepler in the *Library of Useful Knowledge*, and with Sir John Lubbock a little book " On Probability," in the same series.

‡ Described by Sheepshanks in a paper, *M.N.*, **3**, pp. 40–46.

§ According to Robinson, *Proc. Roy. Soc.*, **16**, xlvi. (Obituary of South).

‖ *Publications of the Yerkes Observatory*, **1**, xiv.

SIR JAMES SOUTH
(1785–1867)

To face p. 54.]

South was in the habit of strolling up and down his garden in the evening, shouting his grievances at the top of his voice to some friend, while people from the neighbourhood were regularly enjoying themselves on the other side of the wall by listening to his ravings.   This was certainly easier than to use a large telescope to try to rival the excellent work done by Struve.   Fortunately he did not destroy the object-glass, but presented it to the University of Dublin in 1863, when Lord Rosse was installed as Chancellor of the University.   A few years later it was mounted as an equatoreal at Dunsink.

We have given a rather full account of this affair of South's telescope, although it only indirectly concerned our Society, as the details of it are but little known and, for the sake of Troughton's reputation, deserve to be put in a proper light.   This is the more necessary, as South had a good name as a practical astronomer ; and it should therefore not be forgotten that his charge against Troughton of having failed to make a proper mounting for a 19-foot telescope was not justified.   Of course, his contemporaries knew when he was not to be taken seriously ; so that for instance his grave accusations against the President and Council of the Royal Society in 1830 were quietly ignored.*   We shall close this account of his vagaries by quoting the following characteristic anecdote about him.   Writing in 1836, De Morgan describes how, in the course of a lecture at the Royal Institution " by a starlight Knight," the audience were told " how George III., surrounded by his astronomers, went to Kew to see an occultation, forgoing the stag-hunt which was going on ; how a cloud hid the moon, and how the pious King, without a single murmur against Providence (a point dwelt upon as remarkable), turned the telescope at the hunters, and saw the stag killed between the two horizontal wires." †

2. While the Society was waiting for its Charter, the help of the Council was asked by the Admiralty on a most important subject, the reform of the *Nautical Almanac*.

The Nautical Almanac and Astronomical Ephemeris had appeared since the year 1767.   It was " published by order of

---

* Weld's *Hist. of the R. Soc.*, **2,** 457.   South's pamphlet, *Charges against the President and Council of the Royal Society,*" is dated 1830 November 11. He says he had promised to write a book on the subject, but the unceasing attention which the erection of his large equatoreal had demanded, had prevented it.

† *Memoir of De Morgan*, p. 82 (in a letter to Peacock).   There are two other versions of the story in R. H. Scott's " History of the Kew Observatory," *Proc R. S.*, **39,** 45.   Either the occultation took place in the daytime, or the stag-hunt in the night.

the Commissioners of Longitude," but Maskelyne was responsible for it from the beginning till his death in 1811. During the next seven years nobody in particular seems to have looked after it (though the Astronomer Royal was still supposed to be the editor), and it lost the character for accuracy which the work had hitherto enjoyed. It was said in the House of Commons in 1818 (and could not be denied) that it had become " a bye-word amongst the literati of Europe." This was said during a debate on a bill for reorganising the Board of Longitude. But this Board was, even when thus renovated, an anachronism. The members, who met only four times a year and then only for a very short time, were very numerous and included many whose opinion on questions of astronomy or navigation can hardly have been of much value.* This might not have mattered, if only a suitable person had been selected for the new post of Superintendent of the *Nautical Almanac*. But Thomas Young, who was appointed Secretary to the Board and Superintendent of the *Almanac*, was a perfect stranger to practical astronomy and navigation, however distinguished he was by his discovery of the interference of light and other scientific works, not to speak of his researches on the interpretation of hieroglyphics. Though he did much to retrieve the lost character of the *Almanac* for accuracy, he set his face against every proposal, however moderate, of reform of the publication. One of his arguments, the additional expense, was not worth noticing ; and not much more serious was his objection, that it would confuse sailors to give them a book containing a good deal of information which they did not want. He thought it better to publish predictions of occultations and similar details in the *Journal of the Royal Institution*. But his chief argument was that astronomers had no special claim to be aided in their work at the public expense.†

The call for reform of the *Nautical Almanac* was first voiced by Baily. In the Appendix to the translation of Cagnoli's *Memoir* on determining the figure of the earth (1819), he says that the new Board of Longitude have now the power and the means (£4000 per annum) to enlarge the original plan of the *Nautical Almanac* and undertake other astronomical work. He published in the following year in the *Philosophical Magazine* an ephemeris of the apparent place of the pole star for every day of the years 1820, 1821, and 1822. He next printed for private circulation *Astronomical Tables*

---

* Among them were the Speaker, the Secretary to the Treasury, the Judge of the High Court of Admiralty, etc. The Professors of Astronomy at Oxford and Cambridge were on the Board, but were said not to attend its meetings very regularly.

† Young was thus a precursor of those who fifty years later objected to " the endowment of research."

and Remarks for the year 1822,* a perfect ephemeris except that it does not include the sun and the moon.   As the object was merely to show what an ephemeris ought to contain, the design was taken from Schumacher's *Hülfstafeln* for 1820 and 1821, and much of the contents are borrowed from various sources ; *e.g.*, the list of occultations from Zach's *Correspondence astronomique* (computed for Florence), and the places of the planets from Schumacher's Ephemerides.   Shortly afterwards Baily followed this up by publishing *Remarks on the present defective state of the " Nautical Almanac."*   London, 1822, 72 pp. ; dated May 7.

In this essay Baily first refers to some remarks he had made in the introduction to his recently published Tables, owing to some of these differing from those of a similar kind in the *Nautical Almanac.* These comments had called forth an anonymous " Reply to Mr. Baily's Remarks " in the *Journal of the Royal Institution.*† Baily reprints the whole of this reply and then answers it point by point.   He goes through the four foreign ephemerides and enumerates the articles in them which are *not* in the *Nautical Almanac.*   These were but few and unimportant, so far as the Berlin *Jahrbuch* and the *Connaissance des Temps* went, but the case was very different with the Coimbra and Milan ephemerides, particularly with the former, which Baily pronounced, on the whole, the best pattern for a work of this kind.   His preference of it seems to be mainly due to the innovation of all computations being made with reference to *mean* solar time instead of apparent time.   The Milan ephemeris was specially praised for containing a list of the visible occultations of all stars whose places were given in any catalogue.   Baily also pointed out that the Bureau des Longitudes, established in 1796, more than eighty years after the British one, did not contain any useless members, nor " learned professors, who lived upwards of fifty miles from the place of meeting and consequently seldom attend the Board."

Simultaneously with Baily's pamphlet appeared one by South : *Practical Observations on the Nautical Almanac,* 64 pp., dated 1822 April 15.   He laid particular stress on showing that the *Nautical Almanac* had always contained information which was only of use to astronomers, and that there was therefore good reason for extending the items given.   He compared observations of eclipses of Jupiter's satellites by Beaufoy and himself with the *Nautical Almanac* and the *Connaissance des Temps,* and showed that the data in the latter agree very much better with the observations

* 11 pp. Preface, xxx. pp. Explanation, 72 pp. Tables ; chart of the Pleiades.

† Obviously written by Young himself, as Baily also hints.

than those in the former.   The positions of the planets ought to be given more frequently and more accurately.*

Some of the most pressing needs both of seamen and of astronomers were satisfied by a year-book published by Schumacher for the Danish Hydrographic Office, beginning with the year 1822, entitled *Distances of the four planets Venus, Mars, Jupiter, and Saturn from the Moon, together with their places for every day in the year.*† But Young did nothing ; he had been the teacher of the world as regards the interference of light, but he would brook no interference with his comfortable and not too onerous post as Superintendent of the *Nautical Almanac.* The demand for reform was, however, becoming too strong for him, when in 1827 April, John Herschel, at a meeting of the Board of Longitude, " produced a paper regarding improvements in the *Nautical Almanac.*" Airy, who tells this in his *Autobiography*, adds that Herschel and he were the leaders of the reforming party in the Board, but that Young, the Secretary, resisted change as much as possible. Some slight attempt to satisfy the demand for an enlargement of the *Nautical Almanac* was made by publishing separately a supplement as proposed by Herschel, beginning with the year 1828 ; but it seems to have been issued just at or after the commencement of the year, and could not in any way be considered a satisfactory solution.‡ In the same year, 1828, the Board of Longitude was abolished, but Young remained Superintendent of the *Nautical Almanac.* This year also witnessed the publication of Encke's *Astronomisches Jahrbuch* for 1830, embodying practically all the suggestions made in England.

In 1829 January, Baily issued a second pamphlet, " Further Remarks on the present defective state of the *Nautical Almanac,*

---

* Curiously enough, South (p. 15) expresses his " most earnest wish that astronomers on shore would unite in dismissing mean time altogether from their observatories, knowing as I do, that however suitable to the wants of culinary philosophy, it is only calculated to entail on astronomical observations needless labour, lamentable uncertainty, and, I might almost add, constant error."

† Young made arrangements with Schumacher to have a large number of copies imported ; but only fifty were sold in England, and Young maintained that this proved that practical seamen did not want any information of that kind.

‡ This supplement was published for the years 1828–33.   That for 1831 contains : For every day at apparent noon, mean time, hourly difference, double the sun's daily change of declination, time of semidiameter passing the meridian, sidereal time at mean noon.   For the moon : R.A. and Decl. at time of transit, semidiameter in Sid. T.   For midnight, log. of star constants A, B, C, D.   Hor. parallax and log. dist. of planets for every five days.   Moon-culminating stars.   List of occultations.   This particular supplement was edited by Pond.   We may add that Henderson for some years calculated occultations in advance ; they were at first printed in the *Quarterly Journal of Science*, and from 1829 circulated by the Society in lithographed lists.

to which is added an account of the new astronomical ephemeris published at Berlin." *   In this it is pointed out that the *Nautical Almanac* was never solely intended for sailors ; they do not require to know about the eclipses of Jupiter's satellites or the places of Mercury or Uranus, etc., so that a sixpenny pamphlet would suffice for them.   There are many inaccuracies in the work : invisible occultations or eclipses are marked as visible at Greenwich, or *vice versa* ; mean places of stars are given in one place different from what they are in another ; February 29 of leap year forgotten in the apparent places of stars, etc.

In the same month of 1829 January a Memorandum was (on the 28th) presented to the Chancellor of the Exchequer relative to the expediency of reforming the *Nautical Almanac*.   A month later a motion was made in the House of Commons for the production of papers connected with the late Board of Longitude and the *Nautical Almanac*, and on March 17 these were ordered to be printed.   They are : the Memorandum of January 28, with a copy of the paper read by John Herschel to the Board of Longitude on 1827 April 5 ; also a Report or reply to the Memorandum, by Young, and finally an account of the expenses of the late Board. The Memorandum states that the *Almanac* fell into disrepute after Maskelyne's death ; that there were fifty-eight errors in the volume for 1818 and, singularly enough, precisely the same number of errors in that for 1830 ; that it does not contain the lunar distances from the principal planets, nor any occultations ; that the tables of the sun used by the computers are known to be inaccurate ; that accurate places of all the planets (including the four small ones) should be given for every day, etc.   It is therefore proposed that a new Board of Longitude should be formed.   To all this Young did his best to reply in his " Report " ; but there would be no use in going through his attempts to refute the complaints and deny the necessity of adding to the contents of the *Almanac*.

As the Parliamentary Paper naturally did not contain any refutation of Young's reply to the Memorandum, South thought it incumbent on him to publish a " Refutation of the numerous mis-statements and fallacies contained in a paper presented to the Admiralty by Dr. Thomas Young, etc.," viii + 80 pp.   The preface is dated April 25.   In an appendix are given the Report of 1795 to the French Convention, on the establishment of the Bureau des Longitudes, and the law giving effect to this.   The pamphlet is written in South's usual style, very different from the calm and

* London, 1829 January, 24 pp.   "Extracted from the Appendix to Astronomical Tables and Formulæ."

moderate, but no less convincing style of Baily.* The violence of South was no doubt disapproved by many; thus, Airy writes in his *Autobiography* : "In February and March I have letters from Young about the *Nautical Almanac* : he was unwilling to make any great change, but glad to receive any small assistance. South, who had been keeping up a series of attacks on Young, wrote to me to enquire how I stood in engagements of assistance to Young. I replied that I should assist Young whenever he asked me, and that I disapproved of South's course. The date of the first visitation of the Cambridge Observatory must have been near May 11. I invited South and Baily to my house; South and I were very near quarrelling about the treatment of Young. In a few days after Dr. Young died [on May 10], I applied to Lord Melville for the superintendence of the *Nautical Almanac* : Mr. Croker replied that it devolved legally upon the Astronomer Royal, and on May 30, Pond wrote to ask my assistance when I could give any."

Young's death and Pond's assuming charge of the *Almanac* seem to have caused a lull in the agitation. It was probably thought that the work would at once recover the prestige it had enjoyed in the days of Maskelyne. Anyhow, nothing was done by Pond except, no doubt, to see that the former standard of accuracy was again attained, while he continued the issue of a yearly supplement containing some of the additional information demanded. The call for a more radical reform was, however, aided by the Astronomical Society giving its Gold Medal to Encke in 1830 February for his *Astronomisches Jahrbuch* for 1830. This was the first step of an official character which the Society took in support of the demand for a completely new British ephemeris. South, when presenting the medal (which he did in a moderate and dignified speech), announced that the Admiralty had ordered some additions to the *Almanac* for 1833, and intended to order further additions to that for 1834.

At last the Admiralty made a move in the right direction by addressing a letter to the Council of the Astronomical Society on 1830 July 28. This stated that directions had been given to the Astronomer Royal, who was in temporary charge of the *Nautical Almanac*, to insert certain additions proposed by the late Hydro-

---

* We must give one little specimen. Smyth, when surveying the Mediterranean, was obliged to use the ephemerides of Paris, Milan, Bologna, and Florence on account of the omissions and errors of the *Nautical Almanac* (this Smyth in a letter certifies to be true). But wishing to show civility to a Spanish captain, he presented him with his copies of the *Nautical Almanac* for the current and subsequent years. "Captain Smyth with his foreign ephemerides found his way to England; but there is an awkward story afloat that the Spanish captain has not since been heard of."

grapher, Sir Edward Parry, and "not objected to" by certain members of the Society, to whom they had been communicated. A printed copy of the proposed improvements for 1833 January was sent with the letter, with a request to state whether they were sufficient. On receipt of this, the Council lost no time, but at once appointed a Committee of forty members (including all the members of the Council) to consider the matter. This unwieldy Committee, however, was only nominal, and the work was done by a Sub-Committee consisting of Airy, Babbage, Baily, Beaufort, J. Herschel, Pond, Robinson, South (Chairman), Stratford, and W. Struve. The last mentioned was on a visit to this country, and "devoted a considerable portion of his time to these proceedings."

The "improved" ephemeris for 1833 January submitted to the Society by the Admiralty (all the figures in which were fictitious) was altogether unsatisfactory.* The time of rising and setting of sun and moon were introduced; otherwise the chief alteration was that the place of the moon was given for every three hours instead of for noon and midnight only. The places of the planets were left without change, i.e. that of Mercury was still given for every third day, "the Georgian" for every tenth, all the others for every sixth day, and to 1^m in R.A. and $1'$ in Decl. only.

The Report of the Committee was submitted to the Council and adopted by them on 1830 November 19, when thanks were voted to Baily "for his magnanimous and spirited devotion of his time and talents to the composition and redaction of the Report." It is printed in full in the *Memoirs* (**4,** 449-470), and in the introduction to the *Nautical Almanac* for 1834. The Committee declare that they had constantly kept in view the principal object for which the *Nautical Almanac* was originally formed, viz., the advancement of *nautical* astronomy; but they had also remembered that by a very slight extension of the computations and a few additional articles, the work might be rendered equally useful for all the purposes of practical astronomy.

The first reform demanded was the substitution of mean time everywhere for apparent solar time, though the R.A. and Decl. of the sun and the equation of time should be given both for apparent and mean noon. An additional column to be introduced, giving the M.T. of the transit of the first point of Aries. The use of signs (of the Zodiac) as indicating arcs of 30° to be abolished in expressing longitude. The R.A. and Decl. of the moon to be given for every hour. The time of rising and setting of sun and moon to be omitted. As regards the four principal planets, their places

* The *Nautical Almanac* for 1833, the last one edited by Pond, is in perfect accordance with the plan of the specimen for January.

for every day and their distance from the moon for every third hour were given in Schumacher's Ephemerides, but these were but little known in the British Navy ; it was therefore recommended that these items be given in the *Nautical Almanac*. Also that Mercury and the Georgian be treated in the same way by a " liberal and enlightened Government." Further recommendations included extended information about eclipses and transits of Jupiter's satellites ; the insertion of the list of moon-culminating stars given in the recent Supplements ; the extension of the " elements for computing the principal occultations " into a list of occultations of stars down to the 6th magnitude visible at Greenwich, with elements for predicting occultations of planets and stars to the 5th magnitude visible in some habitable part of the globe. The apparent places of the principal fixed stars to be given for the time of transit and not for noon, and their number to be increased to 100. The several monthly lists of phenomena to be made into one list.

At the meeting of the Council in 1830 December a letter was read from the Admiralty, announcing that the Astronomer Royal had been directed to carry out the suggestions in the Report. But soon afterwards Stratford, Lieutenant, R.N., on half-pay, was appointed Superintendent of the *Nautical Almanac*, and carried out the recommendations of the Committee most thoroughly, beginning with the volume for 1834. Only a very few suggestions were not adopted : ephemerides of the satellites of Uranus, of Encke's comet, and of maxima and minima of Algol. The Society had rendered an important service to astronomy and to navigation by insisting on a thorough reform instead of the half-measures first proposed.

The predictions of occultations commenced by Henderson were continued by Stratford and distributed by the Society till the end of 1833, after which date they appeared in the new *Nautical Almanac*.

There was one desideratum which had not been noticed by the Committee, viz., an ephemeris of the planets for the time of their transit over the meridian of Greenwich. Perhaps they were afraid to ask for too much ; but attention had already been drawn to the utility of an ephemeris of that kind by Sheepshanks, who calculated and printed one for the first six months of 1830. In 1832 November, Stratford offered to provide " a working ephemeris of all the planets at transit," if the Society would pay for paper and printing, which offer was accepted. The same was done for 1834, most of the calculations being done by Mr. Epps, the Assistant Secretary, who undertook the whole of them for 1835. But this was found to take up too much of his time, while the cost of printing (about £30) was rather more than the Society could conveniently afford. Baily, therefore, liberally paid the whole cost of computing and printing for the year 1836. Airy then, in 1836 March, wrote to the Admiralty asking that this Transit Ephemeris might be prepared and printed at the public expense. As he bore the strongest testimony to the value of this publication in fixed observatories, where it saved a great deal of work, the Admiralty at once directed that it should in future form part of the *Nautical Almanac.*

3. The appointment of Stratford to superintend the *Nautical Almanac* obliged him to give up the Secretaryship of the Society, which he had held since 1826. It deserves to be remembered (and is recorded in the obituary notice of him in 1854) that during the five years he was Secretary he had no assistance whatever, so that "the whole routine of the business was conducted by him, from the correction of the proofs of the *Memoirs* to the folding of circulars."   It was in view of his approaching retirement that the Council in 1830 November appointed Mr. James Epps to be Assistant Secretary from December 10, at a salary of £100 a year, and ordered him to attend the meetings of the Council. He was at that time fifty-seven years of age, and though he had not received a regular education, he is said to have acquired a good deal of knowledge of astronomy.   He had published a couple of short papers in the *Memoirs* (vol. **4**) on finding the errors of a transit instrument ; and three others were afterwards printed in vols. **6, 9,** and **11** on similar subjects.   He was also interested in rare, old books,* and was thus in every way well qualified for the post he was to fill. He held the office of Assistant Secretary till 1838 March, when he resigned and removed to Hartwell Observatory, where Dr. Lee had appointed him observer ; but he died in the following year.   He was succeeded in the service of the Society by Mr. John Hartnup, formerly assistant at Lord Wrottesley's Observatory, and employed for some time at Greenwich.   He was engaged at a salary of £80, and held the post till 1843 November, when he became Director of the new Liverpool Observatory.

The growth of the library and many other considerations made it more and more urgent for the Society to find a permanent home in a suitable locality.   They paid fifty guineas a year to the Medico-Chirurgical Society for the use of rooms in that Society's house, 57 Lincoln's Inn Fields.   In 1830 February a Committee was appointed by the Council " to procure apartments."   But they were not easy to find ; and it was therefore fortunate that an influential person came to the rescue. The Duke of Sussex (brother

---

* Among books formerly belonging to Hartwell Observatory there are several rare ones in which Lee has written, " From Mr. Epps's collection."

to the King), who, though not exactly a scientific man, had defeated Sir John Herschel in a contest for the Chair of the Royal Society, offered in 1831 June to assist the R.A.S. to procure rooms in Somerset House in the Strand.   The offer was, of course, gladly accepted, but nothing happened for a long time.   A plan was suggested for taking a house jointly with two other societies ; and on the other hand the Chirurgical Society, who were about to change their residence, wanted to know if the R.A.S. would continue to take rooms with them.   Baily heard in 1832 from the Council of the Royal Society that " there was hope of the matter being accomplished without any further interference on the part of H.R.H."   At last, in 1834 April, the Duke forwarded a letter from the Treasury, stating that there was every disposition to comply with the suggestion, that certain parts of the building lately occupied by the Exchequer offices should be appropriated for the R.A.S., but that the temporary use of them would still be required for a short time.   Finally, Baily as President was able to announce to the Council in the following November that he had taken possession of three rooms on the Mezzanine floor of Somerset House (between the principal and ground floors) and four rooms on the ground floor.   At the Annual Meeting in 1835 February the Council were able to greet the Society in their new home (which they were to occupy for exactly forty years), and to announce that arrangements had been made for the daily attendance of the Assistant Secretary from one till four o'clock.   An additional room * was handed over to the Society in 1836 November.

The library, which had hitherto been " literally inaccessible," now for the first time became of use to the Fellows.   De Morgan had in 1829 offered his services to arrange and catalogue the books and manuscripts belonging to the Society, a task for which his love of books and strong appreciation of the value of accurate bibliography fitted him in an unusual degree.   This work was continued by Mr. Epps, and a catalogue was first published in 1838.   A beginning had already been made towards the valuable collection of manuscripts which now form a very important part of the library.   The original observations of Halley only existed in MS. at the Greenwich Observatory.   In 1832, on the representation of Baily, the Admiralty ordered a copy of these observations to be made and presented to the Society.   This copy was carefully collated with the original, and this interesting series of old observations were thus made more accessible.†   Collated copies of

* " The West room on the Mezzanine floor above the meeting room."

† An Account of Halley's observations was given by Baily in volume 8 of the *Memoirs* (pp. 169–190), and some particulars about his instruments by Rigaud (**9**, 205–227).   Rigaud had in 1832 published Bradley's *Miscellaneous Works*, including many observations not printed before.

Flamsteed's correspondence with Sharp and of Flamsteed's original observations were deposited in the library by Baily in 1834–35.

Another accession to the library of the same kind was the original manuscript of the extensive series of observations of circumpolar stars made by Groombridge. These had been reduced and a star-catalogue prepared from them at the expense of the Admiralty. After 1830 June this work was done by a Mr. Henry Taylor, a brother of the well-known astronomer at Madras, and a son of Pond's First Assistant. He felt aggrieved at the account given of the work in the obituary notice of Groombridge in the Annual Report of 1833 (written by Sheepshanks), though his name was not mentioned in it. His complaint, that statements in the obituary were " totally inaccurate and essentially wrong," was investigated by a Committee, who reported to the Council that his charge was " frivolous and unfounded " ; which report the Council adopted. Upon which Mr. Taylor, deeply offended, resigned his fellowship of the Society. But he would have been much wiser if he had let Sheepshanks alone. For that indefatigable worker, who was now put on his mettle, at once proceeded to make a thorough examination of the reductions and of the printed catalogue, which only wanted the introduction (which was in type) to be printed off in order to be published. This examination led him to find so many errors, that he pronounced the catalogue unfit for publication. At the request of the Admiralty, the matter was next investigated by Airy and Baily, who decided that the errors were of such a nature that no system of cancelling or list of errata could remove them ; so that the catalogue ought to be suppressed. Eventually a new catalogue was prepared under the superintendence of Airy, the main bulk of the reductions being found to have been well done ; and this was published in 1838.

As the Admiralty frequently consulted the Society, it was only a proper recognition of its importance as a scientific body when the President (in 1831) was empowered to nominate five Fellows to serve with him on the Board of Visitors of the Royal Observatory.* The Council also obtained the privilege of distributing a hundred copies of the Greenwich *Observations* (1832).

In the Report on the *Nautical Almanac* the hope had been expressed that a new edition might be issued of the " Tables requisite to be used with the *Nautical Almanac*." In response to this the Admiralty requested the Council to select and arrange tables for a new edition. A large Committee, including several foreign astronomers of note, was appointed in 1831 July. They

* Up to that time the Board consisted of the Council of the Royal Society and a few others nominated by them.

took ten months to prepare their Report, which is printed in volume **5** of the *Memoirs*. They recommended various tables, including a six-figure table of logarithms, which might form a second part to be sold separately. It is stated in the Annual Report of 1833 February that the Admiralty had directed a set of tables to be computed in accordance with the proposal of the Committee, but it does not appear that a new edition was brought out at that time.

Another question, about which the Society was consulted by the Admiralty, was whether, considering the great expense, it was necessary to keep up two observatories in the southern hemisphere, nearly on the same parallel of latitude. These were the Royal Observatory at the Cape and the Paramatta Observatory. The latter had been founded as a private observatory by Sir Thomas Brisbane, and was handed over to the British Government in 1826. The question had already been raised in 1828, when the Royal Society had been consulted and had asked advice from the Astronomical Society. The latter had then declared the two observatories to be necessary; the Cape Observatory as being nearly on the same meridian as the principal observatories of Europe, and that at Paramatta as differing so much in longitude and climate as to be a useful check on the other one. It was pointed out that the southern heavens were very imperfectly known, and that a fixed station on the Australian continent would be of importance for geographical and hydrographic surveys. This opinion was now (1830 December) adhered to, and it was also pointed out that a great deal of money had been spent on the Cape Observatory, which would be wasted if it were given up.

If economy could not be recommended on that occasion, it was duly taken into account in the following year when the Rev. T. J. Hussey asked the Admiralty for £300 or £400 to erect a suitable building to house his instruments at Hayes, in Kent. This the Council could not recommend, though they recognised that Hussey was an active observer, who possessed some valuable instruments.*

The Admiralty was not the only Government Department which showed its confidence in the Council by consulting it in a

---

* Hussey only communicated two short notes to the Society (*M.N.*, **1** and **2**), in the second of which he approved of Bianchini's rotation-period of Venus of twenty-three days. He had a 6½-inch refractor by Fraunhofer, and was the only English observer who made one of the star-maps between ±15° Decl. published by the Berlin Academy. Hussey's Hora XIV. was one of the first of the maps to be issued. He duly entered on this map a star which had been observed by Lalande in 1795; but he did not notice that the star was not there in 1832. It was Neptune! Harding had done the same in 1810. Curiously enough, Hussey was one of the first to consider the possibility of finding the planet which disturbed the motion of Uranus (*cf. Memoirs*, **16**, 387).

way which seems to have fallen into oblivion later on. In 1834 November, Baily, as President, announced that he had during the recess (probably early in August) received a letter from Lord Melbourne, First Lord of the Treasury, requesting that the Council would wait on him in order to recommend a proper person to fill the post of His Majesty's Astronomer at the Edinburgh Observatory, the administration of which had recently been taken over by the Government. As there was not time to call a meeting, Baily and four others had waited on Melbourne, and recommended Henderson ; and this was approved by the Council.* Henderson received the appointment and started work at once ; his observations made up to the end of 1835 were sent to the Council in 1836 to be reported on, as to whether they ought to be printed ; and the same was done the following year, till the Home Office had got to understand that this precaution was unnecessary. The printing of observations seems at that time never to have been undertaken by any public body without the Society being consulted. The East India Company in 1834 was quite willing to allow the Society to pay for the printing of Johnson's Catalogue of 606 southern stars observed by him at St. Helena. Baily, in stating that the catalogue was of a high order of excellence, pointed out that the Society was founded for the collection of the observations of private individuals, not of public institutions, and that their funds were limited. After which the Company agreed to print the catalogue, and it came out in 1835. In the same year the Council was asked to supervise the printing of Maclear's Cape Observations, which request was of course agreed to.

The last occasion on which the aid of the Council was invoked by the Government during this decade was in 1839 March, when the Treasury forwarded a Memorial from a number of people, who had subscribed towards the erection of an observatory near Glasgow, praying for assistance to carry this object into effect. The Treasury requested the Society to give their opinion as to the propriety of complying with this request. The Council recommended this to be done, suggesting, however, that the observations be annually transmitted to the Treasury. This led to the erection of the Glasgow Observatory, which was taken over by the University in 1845.

4. The publications of the Society during this decade bear witness to the rapid rise of astronomy in this country after a long

---

* Thomas Carlyle was a candidate for the post and thought himself ill-used by his friend Jeffrey, then Lord Advocate, " who gave the office to a law-clerk." See *Reminiscences of Thomas Carlyle*, edited by J. A. Froude, London, 1881. As a youth, Henderson had been a writer's clerk.

period of stagnation. As regards astronomy of precision, this rise is more connected with the name of Airy than with any other.

When Airy took charge of the Cambridge Observatory in 1828, he determined at once that the planets were to be observed as often as possible. At Greenwich they had been completely neglected by Maskelyne, who only observed the sun, the moon, and his 36 standard stars, and they had been very little looked after by Pond. Airy now began to observe them regularly at Cambridge, and also showed his interest in them in other ways; by his suggestion that the mass of the moon might be determined by observations of Venus near inferior conjunction,* and by his new determination of the mass of Jupiter.† Having realised the value of regularly continued observations of the major planets, Airy soon saw the importance of getting the Greenwich planetary and lunar observations made since 1750 reduced and compared with the tables. These two great undertakings were not finished till the following decade.

The four minor planets known at that time continued to attract very little or no attention in England, while they were, as in previous years, regularly observed and their orbits computed in Germany. The same was the case with comets; only Halley's comet excited a great deal of interest at its return in 1835. Of researches on planetary perturbations we cannot speak here, since none were published by the Society, but it was during this period that Lubbock published a series of important memoirs on lunar and planetary theory, possessing many novel features.

In order to find a more correct value of the ellipticity of the earth by means of pendulum-observations in high southern latitudes and near the equator, the Admiralty sent out the sloop *Chanticleer* under Commander Henry Foster, R.N., in 1828. On several previous voyages, Foster had made pendulum experiments and taken other observations, for which he received the Copley Medal in 1827. He had served in the *Hecla* on Parry's third Arctic voyage. His work in the *Chanticleer* had nearly been completed when Foster was unfortunately, in 1831 February, drowned in the River Chagres. His observing books and papers were by the Admiralty handed to Baily, who had been partly responsible for Foster's outfit. In addition to two of Kater's invariable pendulums, Foster had taken with him two convertible ones furnished with two knife-edges; these were the property of the Society; they had been designed by Baily, and had been adjusted and tried by him.‡ This led him to investigate all possible

---

* *Memoirs*, **4**, part 2, p. 235 ; *M.N.*, **1**, 140.
† *Memoirs*, **6**, 83 ; **8**, 33 ; **9**, 7 ; **10**, 43 ; *M.N.*, **2**, 171 ; **3**, 36, 113 ; **4**, 25.
‡ Described in *Monthly Notices*, **1**, 78.

sources of error in pendulum experiments, and having become acquainted with Bessel's researches on the correction due to the resistance of the air, he resolved to study the whole subject by new experiments performed under every possible condition of form and material in air and in vacuum.   The result of this most important investigation was published in the *Philosophical Transactions* for 1832.   Baily next reduced Foster's observations and prepared them for publication.   Including London and Greenwich, Foster had visited fourteen stations, the most southerly one being South Shetland, in latitude − 62° 56′.   The compression of the earth deduced from the observations was 1/289·48.   The result found from Sabine's experiments made in 1822–24 from Bahia to Spitsbergen was 1/288·40, but Foster's work was five times more extensive than Sabine's.   The results derived from both these series are in excellent accord with the most recent results of pendulum experiments.   Baily's elaborate report to the Admiralty was printed at the expense of the Government, and forms volume **7** of the Society's *Memoirs* (1834, 378 pp.).

The pendulum being a natural standard of length, it was inevitable that Baily's pendulum observations should lead him to inquire into the question of the British unit of length.   Already in 1830 March the Council resolved that the Society ought to possess a standard scale.   In 1833 the matter was put into Baily's hands, and he had a scale constructed of a novel form, less liable to those sources of error which have so often occurred in instruments of this kind.   The form adopted was that of a cylindrical tube 1·12 inches in exterior diameter and 63 inches long, consisting of three brass tubes drawn one within the other.   The division lines are cut on palladium pins let into the tube.   When in use, the scale is supported on two rollers always placed under the same points.   Three thermometers were let into the tube at equal distances.   The scale was compared with the imperial standard yard preserved at the House of Commons, which was fortunate, since the standard yard was lost in the conflagration of the Parliament building in 1834.   Baily also compared the scale with two copies of the French meter belonging to the Royal Society.   His lengthy " Report on the new Standard Scale of this Society " fills 150 pages of volume **9** of the *Memoirs*.   It includes an interesting history of the standard measures of this country.*

Baily's researches on the figure of the earth naturally led to others on its density.   An accidental remark by De Morgan at the Council table in 1835, that the " Cavendish experiment " ought to

* For the subsequent history of this scale, see *M.N.*, **7**, 55, and **8**, 83. *Cf.* Weld, *History of the Royal Society*, **2**, 267.   It was re-examined by Major MacMahon in 1907 (*M.N.*, **71**, 164).

be repeated, led at once to a Committee being appointed to consider
the practicability of doing so.   A Government grant was obtained,
and Baily undertook the work, which, however, was not finished
and published till 1843.

If from the bodies of the solar system, including the earth, we
pass to the investigation of the elements required for the reduction
of observations, we may begin by mentioning that among the
papers published in the *Memoirs* during the previous decade had
been H. Atkinson's study of the decrease of temperature in the
atmosphere and its effect on refraction (vol. 2).   He followed this
up by a second investigation on the fluctuations of temperature
near the earth's surface, and their effect on the refractions at very
low altitudes.*   Unfortunately the author's death prevented the
completion of his work.   Another paper (of a different kind) on
refraction near the horizon, was one of the results of Henderson's
short stay at the Cape Observatory.   With the mural circle he
observed the apparent zenith distances of stars culminating within
5° of the horizon, on both sides of the zenith.   The result was that
the observations, except in the case of four or five stars, agreed
better with the tables of Ivory than with those of Bessel.†

Of other fundamental determinations we find what is one of
the most important of all, the position of the ecliptic, investigated
by Airy from his Cambridge observations in the years 1833–35.‡
The constant of Nutation was determined by Robinson from
6023 zenith distances of fifteen stars, observed by Pond in the years
1812–35 with the Greenwich mural circle.§

A new value of the lunar parallax was another fruit of Henderson's Cape observations.   He deduced it from observations of
the moon's declination made with the mural circle at the Cape
in 1832 and 1833, combined with corresponding observations
made at Greenwich and Cambridge.‖

But valuable as these results of what might be called Henderson's expedition to the Cape undoubtedly are, they are thrown into
the shade by his great achievement, the first reliable determination
of the annual parallax of a fixed star.   The astronomical world
had grown rather tired of announcements of annual parallax
found from meridian observations.   Brinkley's parallaxes had
been vigorously assailed by Pond ; and though the question
remained in doubt for some years, it was gradually recognised
that they were imaginary.¶   Henderson's paper was laid before

---

* *Memoirs*, **4**, 517–530.   Summary in *M.N.*, **1**, 193.
† *Ibid.*, **10**, 271–282.   ‡ *Ibid.*, **8**, 105 ; **9**, 11 ; **10**, 235.
§ *Ibid.*, **11**, 1–19 ; *M.N.*, **4**, 133.   ‖ *Ibid.*, **10**, 283–294.
¶ Chandler found in 1892 that Brinkley's observations indicated a rotation
of the pole in about a year, and that this would to some extent account for
his strange results.

the Society on 1839 January 11.   By the long delay in reducing his observations of α Centauri he lost the priority of publication, as Bessel had announced the discovery of the parallax of 61 Cygni to the Society two months earlier.

The determinations of astronomical constants referred to in the foregoing, and others published abroad, were urgently required for the reduction of the numerous observations with improved instruments at that time being made.

The Cambridge Observatory was built in 1823–24, but no work of any consequence was done until Airy was given charge of it as Plumian Professor in 1828.   At first he had only a transit instrument and no assistant ; but he fell to work at once, reducing the observations without delay and preparing them for the press, so that the printing actually commenced before the end of the year. The first small volume of Cambridge Observations, 1828, came out in the spring of 1829, soon after an assistant had been appointed. The observations were continued with great regularity, the planets being specially attended to ; but it was not till 1833 January that a mural circle by Troughton & Simms was ready for work.   The observatory was in every way a model institution, and its publications exhibited the reductions to an extent hitherto unknown, while the principle was introduced of not attempting to correct the instrumental errors mechanically, but measuring their amount and applying numerical corrections.

These and other contributions to practical astronomy naturally led to Airy's being appointed Astronomer Royal on Pond's retirement in 1835.   Pond had originally won his reputation by a paper published in the *Philosophical Transactions*, 1806, in which he proved that the serious errors in Maskelyne's declinations of standard stars were due to the great quadrant having become worn at the centre.   At Greenwich, Pond on the whole followed in the footsteps of Maskelyne ; the mural circle ordered by the latter shortly before his death, took the place of the quadrant, and a new transit instrument came into use in 1816.   No improvements were made in the methods of reduction, so that, for instance, Bradley's table of refractions continued to be used long after it had been abandoned as inaccurate everywhere else.   But the observations were certainly better than Maskelyne's, as Pond took great pains to find every possible cause of error.   The greatly increased staff of assistants * also enabled him to multiply the number of single results of any quantity considered to be important.   Towards the end of his life the impression gained ground in London that the Observatory had fallen into a state of disrepute ; and when the appointment was offered to Airy, it was suggested to him that

* There was one assistant when he came and six when he left.

" the whole establishment ought to be cleared out." *   Though Airy acknowledged that " the establishment was in a queer state," he attributed this to Pond's ill-health, to the inefficiency of his first assistant, and to the intolerable amount of business connected with chronometers. This Airy at once got reduced within proper limits, while the first assistant was replaced by a high Cambridge Wrangler (Main), an arrangement continued ever since. The work begun at the Cambridge Observatory was now continued on a larger scale at Greenwich, to the incalculable benefit of astronomy.

At the beginning of this decade the only other observatory in the United Kingdom where useful work was going on and was being published, was that at Armagh, where Robinson had commenced re-observing Bradley's stars in 1827. At Dublin (since the retirement of Brinkley) and at Oxford " grinding the meridian " was going on most steadily and perseveringly, without the slightest thought of reduction or publication. It was no doubt these two observatories which Airy had in mind when he wrote : † " In England an observer conceives that he has done everything when he has made an observation. He thinks that the merely noting the passage of a star over one wire and its bisection by another, is all that can be expected from him ; and that the use of a table of logarithms or anything beyond the very first stage of reduction, ought to be left to others." At Oxford this state of things came to an end in 1839, when Johnson was appointed Radcliffe Observer. Of the work done at the Cape Observatory by Henderson we have already spoken. From 1835, valuable observations were both made and regularly published by him at the Edinburgh Observatory.

As regards instrumental equipment, the transit instrument and the mural circle reigned supreme in British Observatories. Römer's plan of observing both Right Ascension and Declination with one instrument had at last been imitated by Troughton in the transit circle, which he made for Groombridge in 1806.‡ But he never made another, and a few years later he constructed the first mural circle for Greenwich. Why this form of instrument, large and lopsided, should have become such a favourite in this country, though hardly anywhere else, is difficult to explain ; perhaps it was because it was supposed that in order to lessen the effect of division errors the circle would have to be very large.§

* Airy's *Autobiography* (Cambridge, 1896), pp. 109 and 128.
† " Report on the progress of Astronomy during the present Century." *Second Report of the Brit. Assoc.* (1832), p. 184. In a footnote Airy adds that this is, of course, not the character of every English observer.
‡ It had a telescope of 5 feet and a circle 4 feet in diameter.
§ It is, at any rate, something to be thankful for, that the " preposterous " circle (as Newcomb called it) of 8 feet diameter, at Dunsink, was not imitated. It helped to make most of Brinkley's observations useless.

That Reichenbach's transit circle, mounted at the Göttingen Observatory about 1819, was not adopted, is not strange, since the construction was rather weak ; but these faults were remedied in Repsold's form of the instrument.

But however mistaken this policy may have been, there was now everywhere a strong desire to make the utmost of every instrument, and to study and allow for its imperfections. There are several striking examples of this in papers published by the Society ; see, for instance, Sheepshanks' paper on the Cape Mural Circle. From the miscroscope readings at every tenth degree made by Fallows, the first astronomer there, Sheepshanks found that the circle had received some injury, but that the mean of the six miscroscopes was quite to be relied on.* This was afterwards confirmed by Henderson from readings of every 5°.† A very thorough investigation of the Armagh mural circle by Robinson also appeared in volume **9** of the *Memoirs*.

About this time transit instruments were often put to a use which, for some years, threatened to absorb a disproportionate amount of time. This was observing moon-culminating stars to determine the longitude of the observatory, or of some station where corresponding observations were made. Considering the exceedingly rough results obtained, it is strange that this method could remain in favour for some years, even for want of another. But it was not realised that there was no security even in a great number of observations. Thus, Robinson found for the longitude of Armagh, after allowing for irradiation, 26^m $30^s.4$, which he thought could not be more than $0^s.1$ wrong.‡ In reality it was 5^s too small. The determination of difference of longitude by the transport of chronometers, which was first tried between Greenwich and Cambridge in 1828, gradually ousted the moon-culminating stars from fixed observatories.

During most of the time he spent at Greenwich, Pond only observed a small number of standard stars (40 to 60) and published several small catalogues of them. In the Greenwich *Observations* for 1829 he published a catalogue of 720 stars for the epoch of 1830, the largest catalogue based on observations made in England after Bradley's time. Of Johnson's catalogue of 606 southern stars, observed at St. Helena, we have already spoken. The next catalogue to be published in England was one of the Right Ascensions (only) of 1318 stars, observed at Lord Wrottesley's Observatory at Blackheath.§ Mention must also be made of another small star catalogue by an amateur, which, though published in

---

* *Memoirs*, **5**, 325–339, and *M.N.*, **2**, 91–100. The latter is not a mere abstract.

† *Memoirs*, **8**, 141–168.   ‡ *Ibid.*, **4**, 293 *seq.*   § *Ibid.*, **10**, 157–234.

the following decade, was founded on observations made in 1830 and following years. This was a catalogue of 520 stars within 6° of the Ecliptic, observed by Pearson at South Kilworth, Leicestershire, with a transit instrument and a 3-foot-altazimuth.* Of greater importance than these was the catalogue of 726 stars deduced from observations made at the Cambridge Observatory from 1828 to 1835, which appeared in volume **9** of the *Memoirs*. This foreshadowed what might be expected, in the way of perfectly independent catalogues of standard stars, from the Greenwich Observatory under its new director, and was a fitting ending to his work at Cambridge.

If we now turn from the public observatories to those of private observers, we find again one great name which stands pre-eminent ; that of John Herschel. The observations of double stars made by him with his 20-foot reflector at Slough were published in the *Memoirs* of the Society in eight instalments. Six of these † contain the places of the couples found, 3346 in all. The position angles were up to 1828 July 5 merely estimated ; after that date they were measured by a micrometer, but the distances were estimated throughout the whole series. In the first three papers the position angles are expressed according to the notation used by W. Herschel, the parallel being the zero line and the angles counted from 0° to 90° in each quadrant. But in the fourth series (presented 1830 April) Herschel used the notation ever since adopted, having found the old system very liable to introduce errors and confusion. Some members of the Council seem to have been alarmed by this innovation ; and the Committee, to whom the paper had been referred, recommended that the old notation should be adhered to. South was, however, requested to consult with Herschel, and, as an old observer of double stars, he was no doubt easily persuaded of the advantages of the new plan.

The nebulæ and clusters found in the course of Herschel's " sweeps " were formed into a catalogue of 2306 objects for 1830, the single observations being given for each object. About 500 of these objects were recorded for the first time. This catalogue was presented to the Royal Society, and published in the *Philosophical Transactions* for 1833. Our Society's Gold Medal was awarded to Herschel for this work in 1836.

Simultaneously with these observations with the 20-foot reflector, Herschel also made measures of double stars with a refractor of 5-inches aperture and 7 feet focal length, equatoreally mounted. These measures were published in two papers in the *Memoirs*.‡ We may add that another distinguished observer of

---

* *Memoirs*, **15**, 97–127.   † *Ibid.*, **2, 3, 4, 6, 9.**
‡ *Ibid.*, **5**, 13, and **8**, 37.

double stars, Dawes, began his labours on these objects about 1830.*

The brilliant discovery by W. Herschel of binary stars naturally led to attempts being made to calculate their orbits as soon as a sufficiently long arc had been described. The first to do so was Savary in the *Connaissance des Temps* for 1830, and he was soon followed by Encke in the *Berliner Jahrbuch* for 1832. Both their methods were perfect from an analytical point of view ; but J. Herschel considered it a great objection to both, that they required four complete measures. At that time it was assumed that position angles could be measured without much danger of systematic errors, while this was not supposed to be the case with the distances. Herschel therefore rejected the use of distances (except for the determination of the major axis), and found the elements by a happy combination of graphic construction and numerical calculation.†

Herschel's examination of the northern heavens was completed in 1833 May, and as soon as his preparations could be finished (even before all his previous observations were ready for publication) he embarked with his instruments for the Cape of Good Hope, in order to extend to the southern hemisphere the review which his father and he had made of the northern sky. Landing at Capetown in 1834 January, he lost no time in erecting his instruments in a suitable locality about six miles from the town, so that he could begin regular work on March 5. The last " sweep " with the 20-foot reflector was made on 1838 January 22, and thus was brought to a close an undertaking which is unique in the history of science, having been carried out in the course of thirteen years by one individual without any help whatsoever, and entirely at his own cost, including an expedition to a distant part of the earth lasting four years. No wonder that he was honoured in many ways on his return to England in the spring of 1838 ; his scientific friends and admirers gave him a hearty welcome at a festive banquet, before he settled down to the laborious task of preparing for publication the immense number of results of his expedition.

Before Herschel left Slough in the spring of 1840 to spend the remainder of his life in Kent, he had to dismount his father's famous 40-foot telescope, the woodwork of which had become dangerously decayed.‡ This was done in 1839 December, a date which is of importance, as it serves to fix the time of a great advance

* *Memoirs*, **5**, 135, 139 ; **8**, 58, 61.

† *Ibid.*, **5**, 171 ; further applications of the method, **6**, 149. Herschel returned to the subject many years afterwards in volume **18**.

‡ The " Requiem," written by J. Herschel and sung inside the tube on New Year's eve, 1839–40, is printed in Weld's *History of the Royal Society*, **2**, 195, and in the *Astronomische Nachrichten*, No. 405 (Bd. xvii.).

in the then recently discovered art of photography. While the
telescope was yet standing, John Herschel secured a photograph
of it, using a glass negative, which is still in existence and from
which paper prints were successfully made many years later. *
How many glass negatives have been taken since then to depict
the stars and nebulæ, first systematically explored by the two
Herschels ?   It was fitting that what became afterwards a powerful
adjunct to astronomical telescopes should first have been fashioned
by a Herschel, and should first have been directed to the earliest
of modern giant telescopes.

Of private observers with more modest instrumental means
at their disposal, there were as yet very few. Instruments from
the collection formed by the Society were freely lent to such
Fellows as were expected to make good use of them. Among the
earliest donations to this collection were a 4-foot transit instrument
and a small altazimuth, given by the son of Colonel Beaufoy.
These were lent to Captain W. H. Smyth, R.N., who had won a
name by long-continued hydrographic work in the Mediterranean,
and on leaving the sea had settled at Bedford. The altazimuth
was soon exchanged for another (by Troughton) presented to the
Society by Dr. Lee. But the most important instrument in Smyth's
Observatory was an equatoreally mounted refractor by Tulley,
of 5·9 inches aperture and nearly 9 feet focal length, mounted in
1830, and supplied with a clock movement designed by Sheepshanks. With this, Smyth, during the next nine years, measured
hundreds of double stars and examined clusters and many of the
brighter nebulæ. When he had completed these observations,
Smyth parted with his telescope to his friend Dr. Lee, who erected
it in an observatory he had built at Hartwell House, Bucks.†
Here it seems to have been only occasionally used ; but though
never engaged in regular astronomical work, Lee was a generous
patron of science on many occasions and very liberal to our Society,
as we shall see further on.

Another private observatory in the early thirties was that of
Thomas Maclear, at that time a physician at Biggleswade, Bed-
fordshire,‡ where he observed and computed occultations and
other phenomena. But his activity there was not of long duration,
as he was appointed to succeed Henderson at the Cape in 1833.§

* The writer is indebted to the late Sir W. J. Herschel for one of these
prints, mounted in a frame made from the ladder-rungs of John Herschel's
20-foot telescope. The negative is in the South Kensington Museum.

† Hartwell House had been a very well-known place early in the century,
as Louis XVIII. lived there from 1808 to 1814.

‡ Described in *Memoirs*, **6**, 147.

§ It is not a little remarkable that, of four Directors appointed to the Cape
Observatory in fifty years (1830–80), three had already acquired a name as
amateurs.

Of other amateurs active between 1830 and 1840 we must mention George Bishop, whose observatory on the Inner Circle, Regent's Park, London, though started in 1836, belongs more to the next two decades. To Dawes, Hussey, Wrottesley, and Pearson, we have already alluded.   There were not in those days (as there are now) many people deeply interested in astronomy, who, without possessing anything worthy of being called an observatory, yet owned a small telescope or two and got hold of a useful field of work.   Very little, if any, attention was paid to variable stars ; and the study of the surface-markings of the planets was quite neglected.   Of silvered-glass reflectors there were none ; and the possessors of small refractors did not realise that Olbers had never possessed anything bigger than a $3\frac{3}{4}$-inch refractor (or " achromatic," as it would have been called in England), and that Beer's and Mädler's map of the moon and their drawings of the planets were made with a telescope of a similar size.   The English observer with small telescopes had not yet arrived on the scene, but when he did come, his name was to be legion.

But a British amateur astronomer was during this decade hard at work making specula of as large a size as possible.   William, third Earl of Rosse, during this decade succeeded in making mirrors three feet in diameter, first one cast in a number of pieces (mounted in 1835) and afterwards another solid one, mounted in 1839.   His further magnificent success in making a speculum of six feet aperture belongs to the next decade.

Next to John Herschel, the most conspicuous of English non-official astronomers was Francis Baily, of whom it is not too much to say that he was the central figure of our Society during the first twenty-four years of its existence.   In recognition of what the Society owed to him, a number of Fellows subscribed in 1838 and presented a portrait of him to the Society.   It has been shown in the foregoing pages how he, after taking a leading part in the foundation of the Society, endeavoured to encourage amateur observers by the publication of ephemerides and tables, while he, after years of labour, had a principal share in the reform of the national ephemeris.   We have also seen how he was one of the first to grapple successfully with the problem of forming the corrections of a star's place for aberration and nutation into simple formulæ, and how this led him to the formation of the Society's catalogue.   This work on star-places led him also to prepare a new and corrected edition of Mayer's catalogue.   The original observations on which this was founded were published by the Board of Longitude in 1826.   Baily did not reduce them anew, but wherever the positions differed too much from those of Bradley or Piazzi, he searched the observations to find the cause of the discrepancy. The new edition was printed in volume **4** of the *Memoirs*.* Of greater interest was Baily's revised edition of *Flamsteed's British Catalogue of Stars*, chiefly because it was issued together with Flamsteed's correspondence with his former assistant, Abraham Sharp, giving an account of the repeated difficulties and impediments, mainly due to Newton and Halley, which delayed and almost prevented the printing of the *Historia Cœlestis*. At the meeting of the Society on 1833 November 8, Baily gave a preliminary account of the contents of these letters, which was printed in volume **3** of the *Monthly Notices*, 4–10. The whole of the correspondence was then, in 1835, published in Baily's work, *An Account of the Rev. John Flamsteed. . . . To which is added his British Catalogue of Stars, corrected and enlarged* (lxxiii+ 672 pp., 4to).

This publication was objected to in the strongest possible manner by some, who could not believe that Newton, by any possibility, could have been mean or unjust.† Reasonable opponents (like Whewell) were silenced by Baily's reply in the " Supplement " which he printed for private circulation in 1837. It was reserved for an incurable hero-worshipper like Brewster to accuse Baily (long after his death) of " a system of calumny and misrepresentation." ‡ Expressions like these are the more inexcusable, as Brewster, after reading Baily's preliminary paper in the *Monthly Notices*, wrote in 1834 February to suggest to Baily that he might prefix a life of Flamsteed to his edition of the *British Catalogue*, which would afford an excellent opportunity of giving an account of the difference between him and Newton.§ But posterity, which is often more just than contemporaries, has long ago acquitted Baily of the unjust charge brought against him by the blind and uncritical worshipper of Newton.

Baily's further work in the revision of old star-catalogues, from Ptolemy to Hevelius, was completed in the next decade, and published at his own expense as volume **13** of our *Memoirs*. Perhaps we may allude in passing to the phenomenon known as " Baily's beads," a row of luminous points seen by him at the beginning and end of centrality during the annular eclipse of

---

* Auwers made a complete new reduction of the catalogue (Tobias Mayer's *Sternverzeichniss*, Leipzig, 1894). The resulting star-places are vastly superior to those of the former edition.

† When Baily first announced his discovery of the Flamsteed Papers, Ivory called at the Society's rooms to inquire from Epps about their contents and "to express the hope that Mr. Baily was not attacking living persons under the names of Newton and Flamsteed." Ivory passed his life under the impression that secret and unprovoked enemies were at work upon his character. De Morgan, *Budget of Paradoxes*, p. 345.

‡ *Memoirs of Sir Isaac Newton*, **1**, Preface, p. xii.

§ De Morgan, *Newton, His Friend, and his Niece*, p. 106.

1836.* Though seen before, they had not attracted much attention, and there can be little doubt that Baily's lively account of that eclipse did a good deal to excite interest beforehand in the next total eclipse in 1842.

5. During this decade volume 4, part 2, and volumes 5 to 10 of the *Memoirs* were issued, giving excellent evidence of the activity of astronomers in this country, and proving the great value of the *Memoirs* as a medium for the publication of papers on Astronomy. Except volume 7, which contains only Baily's report on Foster's Pendulum Experiments, all the volumes contain many papers of moderate length, which in later years would have been put in the *Monthly Notices*. The Annual Reports of the Council are placed at the end of the volumes, and immediately before them there is (except in 7) a considerable number of observations grouped together under common headings, comets, occultations, eclipses of Jupiter's satellites, moon-culminating stars. The last mentioned were for some years prime favourites, and occupy a good deal of space in print, until it occurred to the Council in 1838 that this was quite unnecessary, since Greenwich, Cambridge, and Edinburgh Observatories published their observations annually. The practice of publishing each volume in two separate parts was discontinued after volume 4, as the papers received steadily increased in number; but as a year or more might often elapse between the reception of a paper and the appearance of the volume in which it was printed, arrangements were made whereby a Fellow, on depositing the estimated value of a volume with the publisher, might be furnished with each sheet as soon as it was printed. But probably very few availed themselves of this privilege.

To some extent this want of rapid publication of results was rendered less harmful by the excellent and fairly detailed summaries of all papers read, which now had become a regular feature of the *Monthly Notices*. These were probably often furnished by the authors, but there can be no doubt that De Morgan, who was one of the Secretaries from 1831-39, deserves a considerable share of the credit of this very useful part of the Society's publications.†️ The *Monthly Notices* had steadily been growing in importance from the first day they began to appear. Started originally to furnish very sketchy " notices " of the proceedings of the Society,

---

* *Memoirs*, 10, 1-42.

† Throughout his life De Morgan continued to be warmly interested in the Society and was a regular attendant at the meetings. This is the more remarkable as he never joined the Royal Society, and described himself as " not a gregarious animal." But he firmly declined the office of President, which he did not think ought to be held by a man who was not an active worker in astronomy.

these summaries gradually became more full; and though as a rule every paper laid before the Society and deemed worthy of publication appeared in the *Memoirs*, there began to be occasional exceptions to this rule, when some short paper would only appear in the *Notices*.* The two first volumes were published by Priestley & Weale; and as the first fourteen numbers (to 1828 November) had run out of print, they were reprinted in 1831, the Society paying half the cost. But in 1834 June the publishers declined to continue the publication of the *Notices*, and volume 2 was therefore at once brought to a close. From that date the *Monthly Notices* were " printed for the Society," and only a number sufficient for distribution to the Fellows were printed.† In the Annual Report of 1840, Fellows were warned that the *Notices* could not be purchased, so that anyone desirous of preserving them should endeavour to prevent their being lost. The result of this reckless anxiety to save a few pounds annually was, that volumes 3, 4, and 5 almost at once became unobtainable; and for the last fifty years or more they have been among the greatest literary rarities. When once the time had passed, when the death of one of the early Fellows would bring a set into the market, they were simply never met with, except when on very rare occasions a long series of volumes might be offered for sale. It is much to be regretted that these old volumes should be so scarce, as not only is much of the information given in them still of value; but the abstracts of papers and the Annual Report afford very pleasant reading.‡

6. It was not only by the number and value of the papers published by the Society during the years 1830–40, that its steadily increasing prosperity was shown, but also by the gradual rise in the number of Fellows. In 1830 February there were 243 Fellows, in 1840, 307. Of these, respectively, 106 and 89 were non-residents, who lived at least 50 miles from London, and, having paid eight guineas, were exempt from annual subscription. In 1831 February an addition to the Bye-laws was passed, putting a stop to the election of non-resident Fellows; but, of course, it took years before the finances of the Society felt the benefit of this change.§ A great

---

* Among these are: Baily's Account of the Flamsteed Papers (3, 4); Sheepshanks' Description of a Clock-movement for Equatoreals (3, 40); Biographical Notes on Halley, by Rigaud (3, 67); Baily's Paper on the Transit of Mercury, 1707 (3, 105); Th. Grubb on Gregorian and Cassegrain Reflectors (3, 177), etc.

† In 1839 May an estimate was received for printing 300 copies. That the volumes from 8 are less scarce is due to Sheepshanks, who for some years had additional copies printed and gave them away (*M.N.*, 16, 91).

‡ The Annual Reports were, however, also printed in the *Memoirs*.

§ The last non-resident Fellow, Admiral Bayfield, elected in 1827, died in 1885 in his ninetieth year.

difficulty with which the Treasurer had to contend, was the considerable amount of arrears of subscriptions which appeared in the accounts every year. In 1830 they amounted to £140, and they nearly always exceeded that sum, reaching £211 in 1839. There seems to have been a disinclination among the members of the Council to take drastic measures to put an end to this nuisance, although the Treasurer or the Finance Committee reported on it several times.

King William IV. had consented to become the Patron of the Society in 1830, and Queen Victoria was pleased to accept the same position in 1837. Two distinguished ladies, Caroline Herschel, the indefatigable assistant of her brother, and Mary Somerville, author of a valuable book on the *Mechanism of the Heavens*, were elected Honorary Members in 1835 February.

Among the benefactors of the Society, John Lee occupies one of the foremost places. He showed his attachment to it in 1831 October by requesting the Council to recommend a candidate for the vacant vicarage of Stone, near Aylesbury, of which he was the patron. The Council " having no candidate before them of known reputation for astronomical requirements," selected one of the applicants. Of more value to the Society was a gift from Lee in 1834 December of £100 as a nucleus of a fund for the benefit of widows or orphans of deceased Fellows. In 1836 April he offered the gift of the advowson of the living of Hartwell, Bucks, which was accepted.* A good beginning was thus made towards the formation of the Society's funded property.

The period 1830–40 was on the whole a quiet period in the history of astronomy. As De Morgan said, " Astronomers had rather given over expecting anything very great in the future : they were inclined to think that nothing was left except to give the existing methods and results additional fulness and accuracy, facility, and neatness." † Considering that the search for a star with an appreciable annual parallax had at last been successful, one would think that many astronomers must have had more faith in the future of their science than De Morgan credited them with.

---

* The deed of this " Voluntary Grant " was not received till 1838 January. Lee had wished it to be stated, that, if the Society should cease to exist, the advowson should go to the Royal Society. But he gave this up owing to legal difficulties, and merely expressed the hope that the Society, while it existed, would never alienate the advowson. But it was ultimately found desirable to dispose of it and the advowson of Stone (given in 1844) to Lee's heir for £700 in 1879.

† De Morgan, *Newton, his Friend and his Niece*, p. 155. Airy, in his *Autobiography* (p. 168), writes in 1845 that the sleep of Astronomy was broken by the discovery of Astræa.

CHAPTER III

THE DECADE 1840–1850. (By R. A. SAMPSON.)

The Society's Rooms.—Throughout this decade the Society occupied its apartments in Somerset House,—"commodious apartments," as Herschel called them, though we should now find them rather narrow. Their position was—

Latitude, 51° 30′ 38″·3 N,
Longitude, 27^s·38 W,

as ascertained by Hartnup in 1843, working with a sextant and pocket chronometer from the terrace, whence he proceeded by an easy triangulation to the meeting rooms. He does not appear to have determined their height above sea level, and it would seem as if these numbers required correction of about $+2''$, $+0^s·25$ respectively, for the Ordnance Survey places the site designated within King's College. The apartments included rooms in which the Assistant Secretary was required to reside (Council Minute, March 1846). During part of the time they included two rooms in the basement. It is noted with satisfaction in the Council Report of 1842, that "Her Majesty's Government has put the Society in possession of two rooms in the basement story of the present building, which have been cleaned out and appropriated for the erection of any apparatus that may be required for pendulum experiments, or for prosecuting any other investigations that may be carried on in such apartments." But congratulation was premature; though the transfer was promised in 1841 June, we find in 1844 February that the Council have to regret that the rooms have not yet been handed over, "although there is no doubt that they are at this moment wholly unoccupied," and it was not till May of the same year that possession was finally obtained. They were immediately used for housing standard copies of United States weights and measures which had been sent from that country, and afterwards for investigations relating to the Standard Yard.

Membership.—The total membership, virtually stationary for the first six years, showed thereafter some rapid accessions : the numbers run—

1840 . . 350	1844 . . 337	1848 . . 364
1841 . . 348	1845 . . 344	1849 . . 388
1842 . . 349	1846 . . 365	1850 . . 412
1843 . . 341	1847 . 365	

These include 36 Associates at the beginning of the epoch and 57 at the end. The accounts show a state of steady, if moderate, prosperity. The invested funds, apart from compounders' fees, increased by £400. There was a sum of floating arrears, which on occasion imposed the unpleasant necessity of expulsion of defaulting Fellows, but " it is gratifying to state that the Society is high among scientific associations as to the promptitude with which its dues are paid."

Presents.—The Society was the recipient of some interesting presents. Among these was Caroline Herschel's telescope, a 7-foot Newtonian reflector, made by her brother and presented by her nephew. A fine altitude and azimuth instrument, constructed by Reichenbach, of Munich, was presented by Admiral Greig, an officer of the Russian Navy. Admiral Greig was a brother-in-law of Mrs. Somerville, and one of the very first members of the Society. He founded the observatory of Nicolajew, and " there is no question that the successful building and endowment of Pulkowa are mainly owing to his care and intelligent guidance." A cast of Chantrey's bust of Mrs. Somerville was presented in 1844. In the same year, Turnor, having acquired some very valuable manuscripts on vellum, containing calendars of the years 1347, 1349, besides planetary tables and other matters, of which the Assistant Secretary, Harris, gave a description in *Monthly Notices*, 1845 January, presented them to the Society with a very graceful letter.* Pearson presented the remaining copies of his *Practical Astronomy*. The generous Mr. Lee presented the advowson of Stone, the second advowson he had made over to the Society. Another interesting gift of his was a portrait of John Middleton, who founded in 1717 the " respectable and useful Society of Mathematicians in Spitalfields," which our Society absorbed in 1846, as related below. The senior surviving member of the Mathematical Society, William Wilson, presented in 1847 twenty-five engraved portraits, which included thirteen of the twenty-three engravings of Newton.

Classed with gifts which show the attachment of Fellows to the Society should be mentioned Baily's payment of the cost of volume **13** of the *Memoirs*, and also Sheepshanks's gift to each of the Fellows of a print of the engraving of the Society's portrait of Baily, " a man whose memory must be an object of almost filial

---

* See Dr. Dreyer's paper "On the Original Form of the Alfonsine Tables," *M.N.*, **80**, 260.

veneration to this Society as long as it preserves its own existence,"
as the Council remark in adverting to the gift (1846 February).
Later, Baily's sister presented the bust which stands in our hall,
" a faithful and charming reproduction " Herschel says of it, " of
features we have so often seen in this place, animated with the pure
love of science and with deep interest in the welfare of this
Society."

Assistant Secretaries.—J. Hartnup, who has been mentioned
above, was one of three men, each of considerable ability, who held
during the decade the office of Assistant Secretary, upon which so
much of the amenity and even effectiveness of the Society depends.
Hartnup was a man of energy, and was appointed in 1843 to the
charge of the observatory which the Mersey Docks and Harbour
Board was about to establish at Liverpool.   He superintended its
equipment—though, as usual, Airy had a large say in this—became
a Fellow in 1844, and contributed frequently to the proceedings, in
particular a description of an improved form of chronometer balance.
He was succeeded by R. Harris, who has also been mentioned above,
" a well-informed and indeed accomplished man," " a student of the
arts of painting and music," of the " propriety of whose manners "
the Astronomer Royal bears witness in recommending him.   He
only held office till 1846, dying of consumption in his 35th year.
J. Williams followed, well-known subsequently for his work on
Chinese Astronomy.   He was a Fellow of the Society at the
time of his appointment, having been admitted as one of the
members of the Mathematical Society, and his knowledge of the
library of that Society, which was in process of examination, and
which proved both valuable and interesting, was immediately
useful.   He resigned his fellowship on his appointment as Assistant
Secretary.

Monthly Notices.—In those days the *Memoirs* were the chief
vehicle of the Society's publications.   The *Monthly Notices* were a
compilation by the Secretaries from such material as was available,
and seldom comprised more than an abstract of an author's communication.   In 1847 the system was reformed.   Thereafter,
Monthly Notices became more full and more defined in form.   Their
contents were to be considered a substantive record of the proceedings of the Society, a portion of its *Memoirs* ; in it alone were printed
such observations or papers as had an immediate interest or were
in a transition state of reduction.   Sheepshanks undertook the
responsible work of editing them.   He inserted an explanatory
note prefacing volume **8**.   The compression and arrangement of
the matter was left in great degree to his discretion.   In arranging
and condensing the *Notices*—for they still were largely abstracts—
he avoided the exercise of any criticism ; it was his object to repre-

sent the views of each contributor in his own language, but he is careful to note that acquiescence with suggestions brought under notice is not to be implied from any silence on his part. Sheepshanks's principal object was to assist in combining and regulating the work of British astronomers, and to publish authentic, original information respecting the progress of astronomy throughout the Empire, and admirably he did his work. Extracts from other journals such as *Comptes Rendus* or *Astronomische Nachrichten* were definitely excluded.

This plan was revised, or rather confirmed, after trial, in subsequent years. In the Report of 1849, we read " in the *Monthly Notices* there is no attempt made to alter the *sense* of any communication ; if it is tolerably ingenious and not positively absurd, the substance is printed in the author's words, compressing the language as much as possible. If a paper appears unworthy of attention (and the Society receives two or three such every session) the nature of the contents is briefly reported to the Council, and a Committee is appointed, to whose judgment the paper is referred." This seems fair-minded, almost to the point of indulgence. " The lucubrations of those authors who treat every science, unknown to themselves, as a new science, and also conceive that astronomy is yet to be *discovered* or rather *guessed*, without geometry, or analysis, or dynamics, are either deposited peacefully in the archives or returned to their authors at the discretion of the Secretary." The Councils of those days, and Sheepshanks, knew as well what was what in astronomy, as any body of men that could be got together.

The Society's Activities.—Perhaps the first impression conveyed, on looking through these early volumes, is the dryness of the material in which our predecessors interested themselves. There was no spectroscopy, no solar physics, no photography. Variables, photometry generally, meteors, parallaxes, systematic proper motions, were all in their infancy. Geometrical and gravitational astronomy had alone attained their full growth and strength. Number after number of *Monthly Notices* is filled mainly with observations and ephemerides of the numerous comets that were discovered, and in the second half of the decade, of the steadily growing family of minor planets. But it was not the view of the Society in those days that the scope was narrow. " It is obvious," wrote the Council in 1845,* " that this is a period of great activity and that all parts of practical astronomy are in full cultivation " ; and again in 1848,† " at the time when this Society obtained its Charter, it was a circumstance not infrequently remarked upon that there was a comparative paucity of great things, accompanied by a constant and

* *Memoirs*, **15**, 407.   † *Ibid.*, **17**, 135.

gradual improvement of routine.   Results of remarkable thought, as well as those of remarkable toil, though not wanting, were not abounding as in the days of William Herschel and Laplace," whereas at the epoch of writing there was an actual plethora.   The Council discerned the signs of their time correctly.   Their concern was that Fellows did not show themselves active enough in sending in those minor communications necessary to keep alive interest at the meetings, which *Monthly Notices* were expressly designed to sift and then to preserve.   The question is touched again in 1847.* " Notwithstanding the signs of activity at home and abroad, and while congratulating our members on the state of astronomical science and the share which this Society has taken in its progress, we may be permitted to remark that a little want of method and perseverance is to be regretted among some of our body.   Several gentlemen possess instruments quite, or nearly, on a par with those of our public observatories, but the actual produce is scanty. It is probable that observations have frequently been made and registered, and even reduced, which have been kept back from a fear of shewing some inexpertness in the minutiæ of practical astronomy. . . .   The friendly advice and criticism of the Members of Council are always at the service of any Fellow, so far, at least, as that knowledge extends."   On another occasion they point out a profitable field for zealous cultivators of astronomy in assisting the production of ephemerides of newly discovered planets and comets, for which up to then the Society was mostly indebted to Hind and Adams, or to communications from Schumacher.

Some of the Fellows.—One would be a poor judge of excellence of character who did not admire men like Baily, Sheepshanks, and De Morgan, to mention no others, and deliberately omitting those whom we now reckon more eminent as astronomers, for the way they guided and shaped the Society.   De Morgan, it is true, has other signal claims to regard.   His personal brilliance, his learning, at once extensive and minute, historical and modern, his hold on the best mathematics of the day, much in advance of his contemporaries, have made his name rather increase than diminish with the intervening decades.   But in his relations to the Council it is his personal side that concern us, that master passion for principle which was more than any reward or success to him.   It finds an interesting expression in the memorial notice of William Frend, his father-in-law. Scientifically, Frend was a bit of a paradoxer, a man who objected to negative quantities, and looked coldly even upon fractions ; but if anyone is interested in De Morgan's point of view, let him read that biography for the way he brings out the beauty and nobility of that simple, self-reliant, truthful character.   Little

* *Memoirs,* **16**, 552.

more than mention can be made here of other greater astronomers in the Society. Herschel during most of this decade was occupied in preparing and publishing his survey of the southern heavens ; he was frequently absent from the meetings and the Council, though several communications show his interest in the Society's affairs, great and small. Airy was at the height of the prestige which he was to retain so long by right of sheer efficiency. His productiveness, the clarity and beautiful form he gave to his numerous contributions, made him invaluable in the meetings. Later, Adams's power made itself felt over his youth and unassuming personality. These men naturally drew the greatest Continental astronomers, Bessel, Hansen, Schumacher, Otto Struve, into close personal relations with the Society by direct communications, by the award of the Medal, by personal visits, or by interesting extracts from private letters read at the meetings. Yet as I see the matter, such men did not constitute the Society, rather they lived upon it. They could have existed as units apart from it ; it was their audience and their stimulus. The Society was the body of ordinary men, trained and judicious enough to appreciate and criticise what was given to them, and to repeat it in some part, humanising the science, bringing with them what bodies of scholars so often lack, the ordinary exacting standards of system and industry learnt in business, and convinced above all, after acquaintance with the world outside, that the rewards of astronomy, such as they were, were well worth their pursuit. Of such men, the " talents were solid and sober, rather than brilliant," and Francis Baily may be taken as their perfected type.

Baily.—Baily died in the year 1844, and at a Special General Meeting convoked to hear a memoir upon his work, it was unanimously resolved that " the Society feels it impossible to express in adequate terms its obligations to its late President "—for he occupied the Chair that year; but what a resolution was too narrow to convey, Herschel's memoir supplies. Herschel's eloquence, often too florid for our present taste, is here sobered by the evident determination not to miss one lineament of his friend or to distort the sterling character he loved so much by any touch of exaggeration. If we were asked to-day how Baily stood as a scientific man, I suppose it would be held he was rather second-rate. His work has not stood very well. His pendulum work contains serious oversights. The Cavendish experiment seemed to have defeated him when a suggestion from Forbes enabled him to complete it. We owe to him in stellar reduction that unfortunate inversion of Bessel's notation, in which while the formulæ are the same, the meaning is different, and which unnecessarily separated British and Continental practice until it was removed from our *Nautical Almanac* a few years ago. But it would be pedantic to measure Baily mainly by such a scale. There is hardly space here to draw a fitting picture of him; reference may be made to Herschel's memorial. Suffice it to say that he was born in 1774, made an extended voyage to the United States, then a somewhat adventurous undertaking; joined a firm of stockbrokers about 1801, and, having accumulated a fortune sufficient for his desires, retired from the Stock Exchange at the age of 51. He was an original member of the Society, and after his retirement from business in 1825, astronomy and the interests of the Society were the undivided objects of his life. He acted as Secretary for the first three years of the Society, prepared all the reports of Council up to the year of his death, and was elected President on four occasions. " To term Mr. Baily a man of brilliant genius or great invention," Herschel writes, " would in effect be doing him wrong." It was his character that left its mark; " its impressiveness was more felt on reflection than on the instant, for it consisted in the absence of all that was obtrusive or imposing, without the possibility of that absence being misconstrued into a deficiency. Equal to every occasion which arose, either in public or private life, yet when not called forth or when others occupied the field, content to be unremarked ; . . . his temper, always equable and cheerful, . . . was a bond of kindness and union to all around him, and inspired an alacrity of spirit into every affair in which the co-operation of others was needed, . . . and brought out the latent warmth of every disposition. Order, method, and regularity are the essence of business, and these qualities pervaded all proceedings in which he took a part, and, indeed, all his habits of life. . . . This was not so much the result of acquired habits as a man of business, as a natural consequence of his practical views, and an emanation of that clear, collected spirit of which even his ordinary handwriting was no uncertain index." One could continue to quote for the pleasure of it, but these extracts are enough. One sees the man he was, and why the Society could find no language to express what it owed to him.

The Society's Outlook.—But to return to the general policy of the Society. While the Society showed such a proper concern to draw all its members, even the less expert, into active participation, it regarded its own duties to astronomy as of the widest and most responsible kind. It was an international exchange and assessor of values, particularly in questions of the award of the Gold Medal. It performed this duty generally in an attitude of judicial impartiality. In regard to the Neptune question, the Council refers to the necessity of " guarding against the undue influence of national feeling," and adds that in this question where a French and an

English claim are mixed, it is not to be regretted that the Society was debarred by the operation of its own laws from deciding between them,—of which more hereafter.

Parallax : Bessel and Henderson.—This careful balance is well exemplified in the case of the Medal awarded to Bessel for the first unquestionable determination of a stellar parallax. This was in 1841. Astronomy had long been plagued with will-o'-the-wisp parallaxes. The annual variation that was looked for opened possibilities of confusion with seasonal changes which might be atmospheric or of various other kinds. The meridian methods employed were not well adapted either for absolute or the highest class of differential determination. Unless a clear, confirmed progression could be shown month by month through the cycle, which could arise from no other cause, suspense and even scepticism was the proper attitude. It has already been mentioned above that Henderson had returned from the Cape in 1833, bringing with him his observations for reduction. He aimed to be, and was, as thorough and careful of instrumental details as Bessel himself, and his discussion of the removal of errors from readings of the Cape mural circle was accepted as a model. In 1839 he produced his discussion of observations of *a Centauri*. The declination, subjected to every test that he could put, agreed with a parallax of about 1″. Yet by common consent, perhaps not excluding Henderson's own, the matter was held as not proven, until Maclear, his successor, should produce a further series that would confirm it. The amount was felt to be large. We now know that the parallax *is* large, the modern accepted value is 0″·76; it is the nearest lucid star yet found ; but Struve had shown, twenty years earlier, that not one out of 27 circumpolar stars whose right ascensions he had examined possessed a parallax of half a second. The confirmation was forthcoming in 1842 ; but it was not reassuring that twenty other stars in Maclear's list showed an average *prima facie* parallax of 0″·3. Not one of these has been confirmed. Henderson remarks : " In a conversation I had with M. Bessel,"—whose friendship was his boast and delight, and whom he consciously took as his model in matters astronomical,—" he expressed his wish that *a Centauri* were observed with a heliometer, or good equatoreal, capable of precise micrometrical measurement ; he said he had doubts of the results derived from meridian instruments. He mentioned the case of Dr. Brinkley's parallaxes, and stated that in his own observatory two excellent meridian circles, placed beside each other, gave at certain seasons places of the pole star that differed from each other ; the reason of which disagreement he had not found out." On the other hand, Bessel's own heliometer measures of 61 *Cygni* left no shadow of doubt that the displacements observed were actually the proportionate projections of the earth's orbit and nothing else. All these points were surveyed critically in the most careful way by Main (*Memoirs*, 12). It was a just estimate of the actual position that awarded the Medal to Bessel in 1841. Henderson has sometimes been blamed for undue caution and delay. This seems a wrong view of the case ; with the means at his disposal, caution and confirmation were an obligation. After his results were confirmed, the Council felt that he too should have recognition. But they missed the right opportunity for action. In 1843 the material was before them, and no name was proposed for the Medal. In 1844 November, Henderson's name was put forward, but in the same month he died. In the same month too, a painful, long and, as it proved, a fatal illness removed Bessel from the scene.

The figure of Bessel, loved and admired, has filled a prominent place in the development of astronomy ; it will continue to do so ; astronomy won him, with its peculiar appeal, in the first flush of his genius and strength, from his clerkship in a merchant's office. He established its foundation as much as it could be given to one man to do. It is surprising that a man with so great an impulse for thoroughness could bring so many works to definite conclusions. For example, he began his studies with the Kœnigsberg heliometer by devoting a paper to the trigonometrical calculation of the field of its object glass. He was known in this country chiefly by his writings, but he visited it in 1842, when he passed a week, along with Jacobi, in Henderson's company at Edinburgh and in the Highlands, and stayed with Herschel, who learnt from him his intention of investigating the errors of Uranus on the hypothesis of an exterior planet.

Fame has given Bessel no more than he earned, but it has done less than justice to Henderson. There can be no thought of comparing the two men together ; Henderson was avowedly a cultivator of the methods of others. " At the outset of his career he was led (probably by the commendation of them in our *Memoirs*) to study attentively the methods of the German astronomers, particularly those of Bessel and Struve, upon whose model he formed his practice, and from which he never departed." I would remark that as astronomy expands, the originator of methods, especially where they involve increase of labour, renders himself more and more ineffectual by his own advances, unless he finds unselfish, able, appreciative imitators to apply his methods far and wide. It needs those qualities, and imagination as well, to see that it is worth doing. Henderson never had a good instrument to work with. It was entirely due to his care

FRANCIS BAILY
(1774–1844)

To face p. 90.]

that any result of value could be derived from observations with the Cape mural circle. After Henderson's time the circle was sent back to England in 1840 to be overhauled, and to Simms' and Airy's great astonishment it turned out that the steel collar was virtually loose upon the pivot; it had never been shrunk on, but was merely attached by soft solder. Yet Henderson undoubtedly exhibited the parallax of *a Centauri* in the measures of zenith distance derived with this instrument. All his other work was equally well judged. At the time of which we write he was living at Edinburgh, but he had formerly spent frequent periods in London, and so was well known to members of Council. Amiable and unobtrusive, he was very modest about his own merits. The biographical notice of his work in 1845 February is written from personal knowledge. "The character of Mr. Henderson as an astronomer stands high, and his name will go down to posterity as an accurate observer, an industrious computer, a skilful manipulator, and an improver of methods in that department to which he devoted himself. . . . Every observation is scrupulously discussed, . . . his processes are fully explained, no labour is evaded, and no circumstance that can affect the accuracy of the final result is passed unnoticed. . . . One of his most distinguishing qualities was sound judgment. He never attempted anything to which his powers were not fully equal; and, as a consequence of this, whatever he did he did well."

Hansen.—In 1842 the Medal was awarded to Hansen for his new methods in planetary and lunar theory. The work had then been applied in outline to the theory of Jupiter and Saturn, and formally, to the moon in the work *Disquisitio Nova.* The great task of calculating the moon's inequalities numerically was still unperformed. It is instructive of the advanced position of gravitational astronomy at that epoch to read Lord Wrottesley's excellent address in making this award. The statement of what Hansen had aimed at and accomplished in his new theory could hardly be improved. The attitude of Hansen to the Society is also interesting. Shortly after this award he found an improvement of his method applicable to the perturbations of very eccentric and highly inclined orbits. He wrote at once to Airy, "I hasten to communicate to you a piece of astronomical intelligence of some importance," and later to Rothman, the Secretary, in similar terms. The Council registered a suitable note of thanks and congratulation on a method "which we are thus far entitled to regard as a most brilliant conquest over one of the residual difficulties of physical astronomy."

The Council had not at that time any practice or unwritten law which restrained it from awarding the Medal to one of its own body. In 1843 it was awarded to Baily, on the completion of the

Cavendish experiment. In 1844 none was awarded. In 1845, Airy, as President, handed it to W. H. Smyth, then Foreign Secretary, for the Bedford Catalogue ; in 1846, Smyth, as President, handed it to Airy for completing the old Greenwich planetary reductions. As regards this award Smyth remarks, " It is, of course, understood, and has always been acted upon, that work, however excellent and useful, does not enter into competition when it only follows the necessary duty of the author. . . . Now the weighty reductions in question come before us as executed at the expense of Her Majesty's Government by the Astronomer Royal. It remains, however, to be added that the undertaking was proposed by that distinguished individual long before his appointment to Greenwich." The distinction seems a just one, but it was hardly necessary to follow, two years later, by an award of a " Testimonial " to Airy for the parallel lunar reductions, which were not actually completed for publication at the time when action was taken.

A Troubled Episode.—We now come to a moment when the Society, from the midst of its harmonious activities, was suddenly precipitated into an acute and bitter controversy, which died down again as rapidly as it had arisen, because no facts were in dispute and there was nothing to controvert. We read in 1846 January, " The addition of a new planet to the Solar System is a fact so interesting and important in astronomy, as to require that the numerous communications of which it has already been the subject should be treated and discussed in the publications of this Society with a greater regard to classification and arrangement than is necessary, or indeed always practicable, in other cases of less prominent interest. . . . It is proposed therefore to give, first, a brief historical notice of its discovery, and of the manner in which the search after it was prosecuted " ; and the next month, among the reports on Observatories, " At Cambridge, the observations of comets and of the new planet have for the present superseded those of double stars." These passages relate to the planet ⑤ Astræa, discovered by Hencke, after a blank of thirty-nine years, the first of a fresh stream that has never since ceased to flow. A little haziness in one's dates might quite well leave the impression that they referred to Neptune. It would have been easily within possibility. In 1841, Adams had " formed a design as soon as possible after taking my degree," of investigating the perturbations of Uranus on the supposition that their unexplained portion was due to the action of an exterior unknown planet. He collected his material, and in the year 1843 he had determined a preliminary place for the body, which was as near the truth as Challis expected his final place to be. The next year he fortified his discussion, and was able to communicate the concluded position in 1845 September to Challis, and in October of the same year to Airy. No one else knew it; Adams told nobody, neither did Challis, neither did Airy, and as far as the Society was concerned nothing happened until the November meeting of the following year.

"J. C. Adams, Esq., B.A., Fellow and Assistant Tutor of St. John's College, Cambridge," was balloted for, and duly elected a Fellow of the Society in 1845 November. Previous to that he had contributed, in 1844 January, a paper on "Elements of the Comet of Faye," communicated by Prof. Challis, where, after computing an elliptic orbit, "the author suggests that the comet may, perhaps, not have been moving long in its present orbit, and that, as in the case of the comet of 1770, we are indebted to the action of *Jupiter* for its present apparition"; and proceeds to show that the planet and comet must have made a near approach so recently as the year 1840. Not a bad first paper, and one that might have drawn some attention to its author. In 1846 April he must have attended the meeting, for he "presented a diagram showing the relative positions of the heads of Beila's comet, and deducing the velocity of the smaller head, finds that its periodic time is 8·48 days longer than the periodic time of the larger."

Le Verrier's Publication.—In 1846 June, Le Verrier published the paper which was the culmination of his investigations upon Uranus, and in which he produced the position of a disturbing planet that would account for the unexplained errors, agreeing in the closest possible way with that which Adams had assigned. It is an extra-ordinary thing that Adams did not seize the occasion to make some announcement of his own parallel, completer, and earlier determination. But Adams was in some respects very immature, and all his life was beset by a peculiar reluctance to performing any ordinary conclusive act, like publishing a paper or even writing a letter. The same cannot be said of Airy. He was immensely struck with Le Verrier's paper. He wrote to him at once to say so, and at the same time put to him his famous poser, his *experimentum crucis*, of the explanation of the errors in the radius vector of Uranus by the same means, which he had also put to Adams, and to which Adams had not sent a reply. The singular thing about this letter is that it did not contain a single word, a hint, that Airy had already had for seven months past, in a pigeon-hole at Greenwich, identically the same explanation of the anomalies of Uranus in considerably greater completeness. It would have interested Le Verrier vastly to know it. It would have prevented the resentment and the charge of disingenuousness which was, not unnaturally, the first feeling which the French expressed on the introduction of Adams's name at a later date. One wonders what Airy proposed to do about Adams ?　He was no novice in reading astronomical documents, and whatever importance he may have attached to an extra verification, he cannot, in my opinion, have been under any doubt of the significance of the brief paper of results which Adams had left with him in 1845 October, or have supposed that the explanation they offered could be explained away.　Yet as for any spontaneous action of his own, then or later, until the force of circumstances had established Adams in a secure position, he seems to have been willing to let Adams's claim and achievement perish unknown.　In excuse for judging so severely Airy's attitude, one must remember the peculiar eminence of his position ; he was the official guardian of British astronomy, and even of science in general, as no one else has ever been ; he deliberately made such a position for himself by cultivating connections at home and abroad, both within and without the borders of his science ; and he was at all times a man of rapid and effective action, never too busy to take up something new.

Anyway, Airy came to the conclusion before the Greenwich Visitation in June, that the planet must be searched for, and that the Northumberland equatoreal, at Cambridge, should be set to the work.　He mentioned it to the Visitors as a matter of necessary division of labour, and referred to Adams's confirmation of Le Verrier.　Herschel was present, and seized the importance of the point at once.　The idea was not a new one.　Hansen and E. Bouvard had canvassed it.　As long before as 1842, Bessel had been Herschel's guest and had talked with him over the errors of Uranus.　He was going to devote attention to it on his return home, and would consider the explanation offered by an exterior planet.　He wrote to Herschel afterwards : " I announce to you (*melde ich Ihnen*) that Uranus is not forgotten," and in fact a young astronomer, Flemming, was engaged by him upon preliminary work at the time of his fatal illness.　Herschel felt no doubt of the existence of the planet, and announced the impending discovery, as far as he felt entitled to do, at the meeting of the British Association in the course of the summer.

Unfortunately the use of the Northumberland telescope meant Challis's direction of operations.　It may be admitted that Challis was a man of no imagination.　The *Athenæum*, in one of its comments on the event, speaks of " the wise men who never believe until the thing is done, the sober men to whom everything that *is* to be is a figment in the brain of a visionary, the practical men who are not quite sure there is a future until it runs by them in the shape of time present."　Challis was one of them.　The search had no attraction for him.　One might suppose he did not want to discover the planet, for when his eye lighted actually upon it in the course of his sweeps, and he made the note " appears to have a disc," he was not sufficiently interested to verify it on the first opportunity. We may agree he did not deserve to find it. It is always a pity when fortune favours the slothful and nerveless. While he continued his slow work, Galle, at Berlin, following Le Verrier's directions, and with the advantage of a good map, found it on the first night of his search, within a degree of the place assigned. Thereupon Herschel wrote to the *Athenæum*, publicly introducing Adams's name for the first time, and immediately after Challis published in the same journal an account of his search.

The Academy.—The French might adopt for themselves the saying, " If it be a sin to covet honour, I am the most offending soul alive." Le Verrier's feat was in direct line with their most illustrious tradition. The direct consequence of Herschel's and Challis's statements was simply not accepted. They were not willing that the newcomer should have any share in the glory. Airy had written to Le Verrier in June and he had said no word of this. Then Airy wrote to Le Verrier again. Adams apparently did exist, but was referred to only in a cryptic way, as though an unpleasing official obligation would compel some perfunctory public reference to him, which Le Verrier must not misinterpret. Airy was surprised that his own name entered at all in the discussions that followed in the sessions of the Academy. " The introduction of my name appears somewhat strange. I have made no public statement whatever regarding the new planet. I have written on it to no foreigner whatever excepting M. Le Verrier himself, and my letters to him (containing some historical statements) were intended to have the most friendly character." In remarking on the course of events in the *Account* read before the Society, he says, " It will be readily understood that I do not [quote this letter] as a testimony to my own sagacity." If he supposed, as he seems to have done, that Adams might be dropped into an oubliette,—of which the history of science has some grim stories,—it showed very little sagacity. The French were not under any delusion about it. They saw their great personal and national achievement assailed, and they were not willing to share it in any degree. There followed an excited session of the Academy, in which Arago's speech may be taken as fairly voicing the feeling. He denied Adams any title whatever to be referred to in connection with the discovery, and personally pledged himself to use no name except *Le Verrier's Planet*. The wilder talk which passed there and outside, especially in the *National* newspaper, was expressly disowned by both Arago and Le Verrier.

The Society's Meeting.—The crisis arose and matured, and the storm burst, between the rising of the Society in June and its reassembly on November 14. At that meeting, with W. H. Smyth in the Chair, after " John Riddle, Esq., Second Master of the Nautical School, Greenwich Hospital, was balloted for and duly elected a Fellow of the Society," the three most remarkable communications were made which the Society can ever expect to receive in one night. The first was Airy's " Account of some circumstances historically connected with the discovery of the Planet exterior to *Uranus*." As the *Athenæum* says, it took the character of a defence of himself by the Astronomer Royal, for not having acted sooner in instituting a search for the planet. The second was Challis's pitiful story,—surely no feebler one was ever told. To do it justice, it is candid. No one would dream of doubting its veracity, for what could induce any man to produce a tale of that complexion ? The third paper was Adams's " Explanation of the observed irregularities in the motion of *Uranus*."

Airy knew how to write. When he was a young man at Cambridge he made it a practice to purify his style by translating, and retranslating back again, to compare with an original model. His *Account* consists almost entirely of letters and extracts from documents connected together by a brief and lucid comment. It strikes me as extraordinarily effective in meeting a tangled situation. Again, no one can possibly doubt its facts. But it leaves one completely at a loss to know why he was so ready to ignore Adams and accept Le Verrier. He never answered this. His radius-vector question was little more than a pretext. A year later, when Otto Struve wrote to him, " L'histoire impartiale, dans l'avenir, citera honorablement et à côté de M. Le Verrier le nom de M. Adams, et reconnaîtra deux individus qui ont découvert, l'un indépendamment de l'autre, la planète au delà de l'Uranus," Airy hastened to endorse the judgment. But by that time Adams was securely established. Adams freely admitted himself to blame for not sending Airy an answer to his question, trivial though he regarded it. But Airy never wrote a word that admitted he had himself wronged Adams by his neglect. It is not unfair to him to say that he preferred for himself the obloquy, that he was ready to exalt the mighty in their seats and to put down the humble and meek.

Adams's *Explanation* is also a remarkable paper. Consider what evidence of immaturity and inexperience he had otherwise shown. Partly by his own constitutional incapacity for action, partly because he was unfortunate in his associates, what everyone then acclaimed the greatest glory of the human mind had been his, and had slipped through his hands, into another's. And he is said at the meeting to have behaved like a bashful boy. But the inves-

John Couch Adams

Engraved by G. J. Stodart from a Photograph by J. E. Mayall.

London, Published by Macmillan & Co.

tigation is as finished, cool, and judicious as a studied report on things wholly remote from the writer. There is not a phrase of complaint, bitterness, or regret. There is hardly a personal word at all, except one generous passage conceding the whole glory to Le Verrier. By the energy of Stratford, who was Superintendent of the *Nautical Almanac,* this powerful memoir was immediately printed and issued as Appendix to the *Nautical Almanac* of 1851, and was thus circulated over the whole world early in January of 1847, before the Council met to settle the award of the Medal. Copies were also circulated by Schumacher with an issue of *Astronomische Nachrichten.*

The Medal.—Hitherto the Society had been, first innocent of the whole affair, then an astonished and excited auditor. Now it had to mark its own judgment upon it. When one is suddenly precipitated before an insoluble problem or into a hopelessly embarrassed situation, what can one do except talk about it ? By the Rules, names proposed for the Medal were submitted at the November meeting and the recipient selected in January. It was decided to propose every name that might conceivably come before the Council. Airy accordingly proposed Le Verrier, Adams, and Challis, thereby contriving to walk down both sides and the middle of the road to show his impartiality. Galle, Argelander, and Hencke were also proposed. Before the January meeting, Adams's memoir was in the hands of the Council, and Le Verrier's name, coming up for a confirmatory vote, failed to receive the three-to-one majority which the Bye-laws required. " It seems to have been thought by several that an award to M. Le Verrier, unaccompanied by another to Mr. Adams, would be drawing a greater distinction between the two than fairly represents the proper inference from facts, and would be an injustice to the latter." Therefore no award was made. " Perhaps there is not one among the Council who does not, more or less, censure the collective body to which he belongs for not adopting a positive course ; while perhaps there are very few indeed who could agree upon any one mode of proceeding." The same Report contains some interesting remarks upon the responsibility of the Council, and the delegation of powers of action to it by the Society. Such a delegation is in fact and practice almost complete, and much greater than the Bye-laws assert. A possible solution considered by the Council was to recommend the General Meeting to suspend the existing Bye-law which required that not more than one Medal should be awarded in any year. In effect this would have been an invitation to the meeting to decide the disputed award. It was contended that " the spirit of the laws would be violated, to the introduction of every disadvantage which those laws were intended to avoid, if

7

a more than usually difficult question were submitted to the Society, of the very kind which the Society had peculiarly delegated to the Council, even in the ordinary and easier cases." Accordingly no such proposal emanated from the Council, but at the meeting several amendments were proposed, of different complexions. A Special General Meeting was assembled the following month to consider the suspension of the Bye-law; but this meeting decided to accept the advice of the Council and to proceed no further in the matter.

> The Council remembers with great satisfaction the amiable tone in which the above differences, more serious than any which had ever prevailed in the Society, were discussed at the meetings ; and they feel assured that in no public body can the prospect of disunion arising out of divided deliberation be smaller than in ours.

It is not now very easy to make out what views were held by what members. Perhaps it may be taken that every possible course found its advocate. It is certain that some of the sternest upholders of Le Verrier as against Adams were found in our own Society and in these islands. The point upon which stress was laid, that Adams had failed to make a technical " publication " of his results, and was thereby disentitled to any share in the credit of the discovery, will strike most people, on review, as extraordinarily narrow and pedantic, even if not wholly a misdirection. The words of Herschel in 1849 upon another occasion, when Bond and Lassell simultaneously had discovered *Hyperion* from opposite sides of the Atlantic, are worth attention. They can hardly fail to have reference to the earlier difficult case of Neptune. "If I am right in the principle that discovery consists in the certain knowledge of a new fact or a new truth, a knowledge grounded upon positive and tangible evidence, as distinct from bare *suspicion* or *surmise* that such a fact exists, or that such a proposition is true—if I am right in assigning as the moment of discovery that moment when the discoverer is first enabled to say to himself, as to a bystander, ' I *am sure* that such is the fact,—and I am sure of it, *for such and such reasons*,' reasons subsequently acquiesced in as valid ones when the discovery comes to be known and acknowledged,—if I say, I am right in this principle (and I can really find no better),"—and so forth. This may not be the easiest principle to follow in making an award according to law, but there is little doubt that it will accord with subsequent settled opinion of what is just.

Testimonials.—But the Society was not yet out of the wood. In the course of the year, no facts being in dispute, opinion settled down to a pretty definite form, from which it has not since much varied. But when November came round, *something* required to be done with regard to the award of the Medal.   Agreement upon a single award was, of course, out of the question.   The solution adopted was remarkable.   Several new titles had arisen in the course of the year that might be held as claims for recognition. " It seemed as if astronomy had exhibited the results of every kind of human aid, and had chosen the year 1847 to show how well she could at once command the highest speculations of mathematical intellect, the laborious perseverance of calculating toil, the discriminating sagacity of the observer, the munificence of mercantile wealth, and the self-devotion of the voluntary exile." When our grandfathers indulge in a sentence like that, we can only bow and stand aside.   They decided not to suspend the Bye-law which limited them to one Medal, but to award Testimonials, as many as the occasion demanded.   Everybody who was proposed got one,—twelve in all,—Hansen, Hencke, Herschel, Hind, Lubbock, Le Verrier, Adams, Argelander, Bishop, Airy, Everest, Weisse.   The proposals were made by Airy and De Morgan, except Weisse's name, which Galloway added.   Hencke and Hind had each by that time two minor planets to their credit.   Bishop got it for main- taining his observatory.   Airy for the Greenwich lunar reductions (not then completed).   Hansen had received the Medal as recently as 1842, and Airy in 1846, Herschel, Hind, Airy, and Adams were on the Council.   So with this remarkable procession of talent the troubled incident passes out of our annals.   The action of the Council cut the knot, at the expense of prematurely rewarding some and unnecessarily rewarding others, robbing the gift of its one value, rarity and distinction, offending against good taste by rewarding several of its own members, and depleting future years of many of their best candidates.

But immediately after the awards fall into their old excellent habit, with Lassell in 1849 for his 24-inch speculum and discoveries of satellites with it, and Otto Struve in 1850.

The Mathematical Society, 1717.—One of the most interesting events of this decade was the absorption in 1846 of the Society of Mathematicians of Spitalfields.   From the earliest days of our Society our *Memoirs* were presented to the Mathematical Society, for the latter was our senior by more than a century.   It was founded in 1717 by one John Middleton, whose portrait we possess.   The portrait shows a man of benevolent and practical appearance, holding a geometrical diagram ; a ship under sail is in the background.   It is conjectured he may have been a mariner, who gave instruction in navigation.   I give here what is recorded of this curious Society ; one would wish to know more, but its early activities can only be guessed from the library it collected.   The first Minute-books in existence date from 1800, and reveal the Society suddenly confronted by a most real and substantial danger. A gang of informers had laid an information against some of the members ;—its tenour is not quoted, but one of the members, Gompertz, afterwards told De Morgan that the charge was for taking money for an unlicensed entertainment,—being a philosophical lecture. The members, about forty or fifty in number, raised in a few days a voluntary guarantee fund of £254, one of them undertook without payment the professional part of the defence, and the informers were beaten off at a cost of about £43 in expenses. These vermin do not seem to have been liable to any punishment for their baseless charge, for they went away threatening to return to the attack. But though the charge was rebutted, it appears to have carried with it a certain scandal, for we find it noted that the " produce of the lectures delivered in 1799–1800 had been very materially diminished by the effect of the information lodged against several of the members by the Gang of Informers, who have occasioned so much trouble and expense to the Society during the past year." The Society, which was very straightforward and democratic in its constitution, levied a charge of threepence a week on each member, calculating that in about eighteen months that would repay what some of them had advanced to meet the expenses. Some of our Minutes refer to the "respectability" of the Mathematical Society. That it was fully "respectable" in the modern sense may be conceded. Among the motions in its Minute-books is one " that every member who may so far forget himself in the warmth of debate as to threaten or offer personal violence to any other member, be liable to be expelled." But that, after all, is an evidence of the desire to keep good order. In 1802 an application was made and agreed to, to hire the premises, when not in use, for the purposes of a Sunday School.

The Society was united on a democratic and social footing. There is little or no academical trace in its membership. De Morgan says it consisted originally of Spitalfields weavers, but of these there is no trace after 1800. The new admissions seem to have been, indifferently, tradesmen and professional men from the immediate neighbourhood. Thus we find several surgeons and attorneys, a wine merchant, a mason, a tinplate worker, a hairdresser, a chemist and druggist, a mathematical-instrument maker, a watch-case maker, a watchmaker, a " plaisterer," a painter, a dyer, an Exchange broker, a glass-cutter, a schoolmaster, —apparently any ingenious man of any occupation was welcome. Two intriguing occupations are " Gentleman of the Prerogative Office," and " Galenical Operator in Apothecaries' Hall." There are no distinguished names.

Its Rules.—It was one of the Rules that it is the duty of every member to answer the best way he is able any mathematical or philosophical question that may be asked of him. Failing which he was fined twopence. These discussions were sped with the help of pipes and porter. After the unpleasant shock of the Information the use of these adjuncts to philosophy was stopped, as an institution, in 1801, and left to the private taste of members ; but the change was not a success, " the small number attending on Saturday evenings arises in great measure from their not drinking in common," and the old custom was restored. Another Standing Rule provided that " every member shall in rotation give a lecture or perform some experiment on Saturday evenings." As a reward for these lectures, medals were given. There should be some of these in existence, as prior to 1801 they were " given with too much facility " ; at that date it was decreed that a larger medal of 2 oz. of silver, and a smaller one of 1 oz., should be balloted for ; the scale for voting was elaborate,—5=excellent, 4=very good, 3=good, 2=tolerable, 1=scarcely tolerable, 0=very indifferent. The medals were to be presented with due ceremony on Newton's birth-night. No doubt this habit of holding lectures trained and brought out a number of capable lecturers, for we find an annual course, open to the public for a fee, in existence at the beginning of this time. These lectures were on what would be called Natural Philosophy, and were illustrated by experiments.

Its Lectures.—The purchase of the necessary apparatus resulted in the collection of a considerable store by the Society. Their subjects and number varied a little from year to year. One of the most ambitious courses was held in 1821, when 5 different lecturers delivered between them 22 lectures in all—3 on Mechanics, 2 on Hydrostatics, 2 on Pneumatics, 2 on Optics, 3 on Astronomy, 6 on Chemistry, 1 on Magnetism, 2 on Electricity, and 1 on Galvanism. The charge for admission to a single lecture was 1s., and to the course 15s. This course resulted in a net profit of £67, 17s. 3d. to the Society. Similar substantial profits were made in several years, and the library benefited greatly in consequence. But a time came when the small body of which the Society always consisted, found itself unequal to the task. It is reported in 1825 that " great difficulty was found in procuring members to give the lectures," and they were reminded of the Standing Rule. But lectures given in obedience to a rule proved unattractive, and the Society experienced for the first time a loss, amounting to £7, 6s. 8d. Thereafter the lectures seem to have been dropped. Indeed, the future of the Society itself soon became a cause of disquiet. It was reported in 1829 that the income of the Society for the past five years had been respectively £146, £151, £135, £140, £105. No new members joined. A valuation was made of the movable property ; it seems of an optimistic character. The total was £2649, of which " printed books, maps, and prints " accounted for £1565, and " instruments, natural history, and antiquities," £664. One may doubt the value of many of these curiosities ; in increasing frequency the Society's collection seems to have become a dumping ground for the domestic superfluities of its members,—skins of penguins, a skeleton of a Borneo monkey, a head of an antelope, stand outside its pristine functions.

Its Decline.—But it struggled on. In 1841 the number of members was 30. In 1843 they removed from Crispin Street to 9 Devonshire Street, and sold instruments to the value of £70. In 1844 a motion was made that the library of this Society be given to the library of the City of London, upon condition that "members of this Society be permitted to use the same during their lives," but no action was taken. A few months later it was proposed that all the books, pictures, and other chattels be drawn for by the members by lot.

Absorption.—At this time the Society consisted of nineteen members, of whom three—Dr. John Lee, Benjamin Gompertz, and J. J. Downes—were also Fellows of our Society. Lee, who was eminently a good Fellow, took the situation in hand. He talked it over with our Council, and wrote a letter on 1845 May 10 to the Mathematical Society. He says, " A meeting of the Council of the Royal Astronomical Society took place yesterday, and I brought forward the suggestions contained in your recent letters to me relating to the venerable Mathematical Society of London, and the Council were unanimous [in regretting] that this ancient Society of 130 years' standing should be on the eve of dissolution and decline. The members of the Council were also, I believe, unanimous that if the nineteen surviving members of the Mathematical Society should in their liberality and public spirit wish to keep the mathematical, and astronomical and philosophical portions of their valuable library together, and should kindly and considerately offer to present it to the Royal Astronomical Society, that the Council of the latter would not only be grateful to them for this act of judicious benevolence, but would be willing to elect all the members of the Mathematical Society members for life of the Royal Astronomical Society. . . ." It was arranged that a visitation should take place by the President and Secretaries, Smyth, De Morgan, and Galloway. Lee concludes his letter : " I hope that this matter will terminate successfully and beneficially for both these noble-minded Societies."

Surely no angel ever beckoned more beautifully a spent soul to euthanasia. The terms proposed were carried out. The books were transferred, and the instruments, pictures, and curiosities divided by lot. Some of the most interesting were afterwards presented to our Society, Lee making a gift of the portrait of John Middleton, referred to above, and Mr. William Wilson, the senior member of the old Society, presenting an album of 25 portraits, including 13 out of 23 known prints of Newton.

Its Library.—The history of the Mathematical Society, so far as we can glean it, is certainly a very curious episode in British science. In scientific value it is unfortunately nil, apart from the collection of a library, but it illustrates a national capacity and inclination, which our own Society and a host of others also exemplify, for forming circles with disinterested aims and keeping them in permanent being. The library consisted of upwards of 2500 volumes. The collected catalogue of them is not now available. When they were incorporated with our books, many volumes were described as of an unusual, some of a rare, character ; mostly mathematics and chemistry of the seventeenth and eighteenth centuries. Some of the lists of purchases, and a few dozen of the volumes, handled at random, reveal, for example, Clairaut, *Figure de la Terre* ; *Commercium Epistolicum* ; Euler, *Theoria Motus Lunæ* ; Boscovich, *Opuscula* ; Taylor, *Sexagesimal Table* ; Burckhardt's *Tables* ; Delambre's *Tables* ; Flamsteed, *Historia Cœlestis* and *Atlas* ; *Histoire de l'Académie des Sciences*, 113 vols., 12mo ; *Euclidis Elementorum Libri XV Grœcé et Latiné*, Paris, 1573, " Liber rarissimus " ; Bernoulli, *Doctrine of Permutations*, . . . edited by Francis Maseres, and presented by him, 1795 : Euler, *Institutiones Calculi Integralis* ; Arbogast, *Calcul des Dérivations* ; Robert Boyle, *Tracts*, 1672 ; d'Alembert and Condorcet, *Nouvelles Expériences sur la Résistance des Fluides*, 1777 ; Desaguliers, *Course of Experimental Philosophy*, 1763. There are besides the current English mathematical textbooks of the period, Wood, Bonnycastle, etc. ; Newton's commentators and diluters figure pretty strongly. From these specimens it may be accepted that, what with purchases and what with gifts, the library was a fairly enterprising collection for its period.

The pathetic thing is, that with all its long life and good intentions the old Mathematical Society never became a place where men really cultivated mathematics. We live in a country where a George Green, a Dalton, a Faraday, though rare, of course, are rather characteristic than singular. On a lower scale one Fellow, Professor Wallace, of Edinburgh, who died in this period, was a bookbinder's apprentice, who began his own education on the books of science which passed through his hands. The mathematics he wrote, about 1800, though no great affair, are quite in line with the developments of analysis which Euler had taught.

The Small Achievement.—But we were still in the period of that long blight on English mathematics, following Newton's flights, which were nowhere more wonderful than in pure mathematics. Whether from anti-Continental feeling, or from national obstinacy, or the idea that the tangible methods of geometry offered the sure slow road to truth, our workers crippled themselves. The period of Euler, d'Alembert, Legendre, Lagrange, Laplace, Abel, Cauchy, Jacobi, Gauss, passed over their heads almost without attracting their remark. Oxford and Cambridge were no exception. Thomas Young was not to the taste of Professor Vince : " What do you think of a man writing on mechanics who does not understand the principle of the coach-wheel ? " Professor Vince asked. I have often wondered what is the principle of the coach-wheel. No doubt it is enshrined in many a problem paper of the period ; it is sufficient, however, that it excluded from that gentleman's field of view matters we have come to think more important.

The old Mathematical Society was no worse than this ; the pathetic thing is that though unhampered by interest or tradition, it was no better. The most celebrated names that it can claim are Dollond and Thomas Simpson ; it had no luck in drawing to its hearth any spark of native genius, or even in forming itself a centre for understanding the wonderful structure which mathematics had become.

Other times, other modes. We can close on a different note. We now have as guests in the Society's rooms another London Mathematical Society, whose members are more able to criticise astronomers for backward methods and deficient analysis in mathematics than they are likely to lay themselves open to that charge.

American Astronomers.—One of the features of this decade is the definite entry of American Observatories into the Society's field of view. In 1847 the Council writes : " It has often been a matter of regret, and sometimes a ground for reproach, that the vast country of the United States has shown so little interest in the science of astronomy. This apathy, at any rate, exists no longer. Observatories fully equipped have been erected at many places." And they proceed to instance the equipment of the Naval Observatory, Washington, and that of Cambridge University, U.S. For the latter, " an equatoreal instrument, similar in size and mounting to that of Pulkowa, is now constructing by Merz, of Munich, we presume at the cost of the state of Massachusetts." But in this surmise they were wrong ; the 15-inch equatoreal of Harvard College Observatory was provided at the cost of the College and by private subscriptions. It was erected in 1847 and immediately established itself under W. C. Bond, running a curious race of rivalry with

Lassell's 24-inch speculum, and discovering *Hyperion*, Saturn's 8th satellite, simultaneously with him. But the Council says, "the most remarkable foundation, and that which does most credit to the energy of Americans, is the erection of an equatoreal telescope of 12 inches aperture (by Merz, of Munich) at the thriving and spirited town of Cincinnati. The funds were obtained by subscription from individual citizens on the solicitation of Mr. O. M. Mitchel, who has undertaken the charge of the Observatory. . . . If as much skill be displayed in working the instrument as has been shown in procuring it, the Observatory of Cincinnati will soon be celebrated among its compeers."

W. C. Bond was made an Associate in 1849 January, the first American added to the list, and, as Herschel said, " not long to be the only one of his countrymen by whom that honour is enjoyed." Next year the Council announced that they felt it their duty to increase the list of Associates by the addition of names from the United States, adding that the very great impulse given to astronomy in that part of the world within the last few years will ultimately demand many acknowledgments of the same kind. The names added were B. Peirce, A. D. Bache, O. M. Mitchel, and S. C. Walker.   Peirce was at the time best known for his celebrated dilemma, that Neptune, as discovered, was not the planet of prediction, since its perturbations of Uranus were of a very different character from those contemplated, and was, moreover, explicitly excluded because it did not conform to the limits of distance which Le Verrier had laid down. Bache was Director of the U.S. Coast Survey, and had communicated their first experiences with an electro-chronograph. Airy seized the idea at once, " apparently suggested by the obvious practicability of applying the galvanic telegraph (so extensively used in America) to the determination of differences of longitude," and made plans for the chronograph, governed by a conical pendulum, which has run so long at Greenwich. He remarks : " The Americans of the United States, although late in the field of astronomical enterprise, have now taken up that science with their characteristic energy, and have already shown their ability to instruct their former masters." Among other evidences of this energy which are forthcoming in the period under review are an expedition to Chile for the purpose of observing Venus and Mars and determining the solar parallax by comparison with concurrent European observations, and the discovery of a comet by Miss Maria Mitchel, of Nantucket.

The Cape.—The Society was by this time a convenient medium of exchange for astronomical ideas from all over the world. The foregoing passages by no means exhaust the references to the United States. Another source from which interesting news and reports frequently arrived was the Cape, and that from two connections.   First, Herschel had only recently returned from his expedition, in which he had set up his 20-foot reflector for a survey of the southern heavens.   The strong personal charm and influence his presence never failed to exert is apparent. In 1844, Maclear sent an account of the erection of a memorial obelisk: " Sir John Herschel, during his residence at the Cape, was President of the South African Literary and Scientific Institution. When he was about to leave the Colony, the members expressed a desire to present him with some token of remembrance ; and at a full meeting a few days before his departure, a gold medal was presented with the impress of the Institution on one side and a suitable inscription on the reverse.   The feelings excited on that interesting occasion strongly evinced how much the members regretted the loss of the President, and their admiration of one whose talents place him so far above ordinary men, and whose private life was a pattern of every domestic virtue."   Accordingly an obelisk was erected, on the site of his telescope, and various memorials were immured below.

Herschel's work at the Cape was, as is well known, independent of the Cape Observatory.   The reports of that Observatory cover at least three matters of interest.   Henderson had returned from there in 1833.   Maclear's confirmation and continuation of his parallax work has been mentioned elsewhere.   Maclear was himself engaged for most of the period in a very laborious and punishing geodetical expedition in repetition of Lacaille's meridian arc.   The matter was complicated by questions of local deviations of the vertical at the two ends, the arc beginning north of the Cape mountains and ending to the south of the Kamiesberg.   At the latter the country was absolutely wild, unknown even to the natives, waterless and exposed.   The work was courageously completed, but Maclear's health suffered severely.   The third matter was the publication of the history of the foundation of the Observatory, and the first observations made by Fearon Fallows there.   Here Airy's energy in reducing old observations came to aid.   He discussed the material which Fallows had accumulated from 1829 to his death in 1831, and made it available, for what it was worth.

India.—Madras Observatory may also be mentioned as contributing to the flow of news ; here T. G. Taylor was at work until his death in 1848.   Less familiar, and therefore perhaps more welcome, Trevandrum on the opposite coast.   Caldecott was astronomer to the Maharajah of Travancore.   Besides numerous observations of comets, Caldecott shows the true spirit in his observations of the solar eclipse of 1843 December.   Having ascertained that the eclipse would be total for a brief space at an accessible point near Tellicherry, on the coast, he proceeded there with his instruments, ascended a river to its source and partly surveyed it. He obtained his observations.

In the case of comets, especially the great comet of 1843, observations of varied rank and value flowed in by correspondence from all quarters of the world, and were recorded by the Society.

Foreign Astronomers.—Foreign astronomers made hardly less use of the Society for announcements. Schumacher, in particular, was indefatigable in sending ephemerides and other news of interest. And in return, when the troubles of 1848 brought Schumacher's position at Altona into jeopardy, arising from a rebellion in the Duchies of Slesvig and Holstein against the King of Denmark, all the astronomers of Europe used what influence they had to support him; the Society sent a deputation to Lord Palmerston, by whom it was sympathetically received, and the case immediately represented to the Danish Government.

Instrumental Advances.—The interest of the Society in instrumental advances was keen, and with few exceptions the lines approved have stood the test of experience. In 1843, Simms described his new dividing engine, which was self-acting. Airy successively described his plans for the new transit circle, the reflex zenith tube, and the chronograph. He also described with care and fulness the methods of casting and grinding specula, of Lord Rosse and of Lassell. Lord Rosse contributes some interesting remarks on the mounting of a great mirror, which are still to the point. Lassell was a frequent contributor as well as an indefatigable observer, and his 24-inch speculum, with its equatoreal mounting, must have been an unusually fine piece of work, as evidenced by the discovery of Hyperion, of two of the lost satellites of Uranus, and of the satellite of Neptune. Some of the subjects talked about seem to us strangely familiar, though at the time they were mere talk; for example, "The advantage of large specula and elevated positions," by Piazzi Smyth. The Neilgherry Hills was the site he suggested, that is to say, not so far from the site of Kodaikánal.

Equally assiduous were Fellows in studying improvements in methods of using the instruments and the minutiæ of reading upon which refinements depend. Sheepshanks was one of the most expert. The many entries regarding the standard yard offer an illustration. The Society possessed, and still possesses, a copy of the national standard. The original was destroyed in the conflagration at the Houses of Parliament in 1834, and at the request of the Government the Society's copy was lent for the purpose of constructing by comparison a new standard. The work went on slowly; all the difficulties of this branch of metrology had to be found out one by one, composition, support, illumination, personality ; and the matter passes out of the decade unconcluded.

Glass.—The Council in 1846 February, " cannot but mention what is one of the most remarkable events of the year, though perhaps no one of the parties concerned in it gave our science a thought. They allude to the repeal of the excise duty on glass, which might be called with perfect truth an astronomical window tax. The regulations up to then rendered experiments for the improvement of optical glass almost impracticable, and a great deal too expensive. It may now be confidently hoped that in a few years our country will not be obliged to admit that we are surpassed by foreigners in this particular."

The difficulty was very real. Simms was unable to secure the flint glass to make the Liverpool equatoreal (8-inch), and the whole was ordered from Merz, of Munich. The same artist had supplied the 15″ O.G. for Cambridge, U.S., and 12″ for Cincinnati. Later on, however, Simms succeeded in making an 8″ O.G. (12′ 6″ focus) for the new Greenwich transit circle, which satisfied Airy's tests— it separated η Coronæ (0″·6) but failed at γ Coronæ (0″·4). The price was £275 ; where the glass was obtained is not stated. Soon after he was able to announce that the difficulty was at an end (1849 April)—" the firm of Chance & Co., of Birmingham, with the assistance of a foreign artist, have succeeded in manufacturing flint glass for optical purposes by no means inferior, so far as my trials enable me to judge, to the very best that was formerly prepared by the elder Guinand." This foreign artist was H. Bontemps, whom the troubles in Europe in 1848 had driven from Choisy-le-Roi, and to whom the younger Guinand had communicated the method that had made his father's, and Fraunhofer's, and Merz and Mahler's fame, of stirring the melted glass till it could be stirred no longer, and then chilling the pot somewhat quickly, by which the melted mass split itself into blocks, each sensibly homogeneous.

Eclipses.—There are numerous eclipse observations within our period ; mostly they are filled up with the tedium of Baily's Beads, but there is a notable exception. In 1842 a total eclipse of the sun took place in North Italy and Austria. Baily and Airy saw it, the former from Pavia, the latter near Turin. Both sent vivid accounts ; Airy's in particular is tremendous, and quite outdoes the reality of most experiences. But what matters is, that both record and draw the pink " protuberances " then noted for the first time.

Solar Physics.—There was no physics of the sun in this period ; Schwabe's announcement of the sun-spot period in 1844 in *Astr. Nach.* seemingly attracted no one's attention, or at least belief, and it is curious to find Herschel in 1847, while exhibiting a month's hourly drawings by Griesbach, recommending, without any mention of Schwabe, as "highly desirable to secure an unbroken series of drawings exhibiting a continuous view of the changes in the sun's surface for every day in every year in future, and as near an approach to it in past years as can now be recovered.   It seems high time that some attempt of the kind should be made on a systematic and regular plan, as the only probably effectual means of arriving at a knowledge of the laws which govern these mysterious phenomena, and the periods, if any, which they observe in their formation, and thence of elucidating the nature of the sun itself." He goes on to recommend the Society to start what might have been the beginning of a Solar Research Union (November 1847). But that was for another epoch ; and it is time now that 1840–50 gave place to 1850–60.

CHAPTER IV

THE DECADE 1850–1860. (By E. H. Grove-Hills)

THE period that we now enter upon is a profoundly interesting one, not only as considered from the view-point of astronomical progress, but as marking a fundamental change in the aspect and methods of all the physical sciences. It was indeed in a very real sense a transition period, when the old problems and the old modes of attack were either solved or exhausted, and when new questions, new and powerful instruments of research and new resources were being rapidly disclosed and developed.

It will be hardly necessary to disclaim any intention of an attempt to write a history of the progress of astronomy during these ten years ; such would lie outside our scope, and space would not permit us to do it justice ; still less shall we venture upon an outline of the general progress of physical science during the period. It does, however, seem desirable, and in fact necessary, that before we enter upon our real theme, the history of the Royal Astronomical Society, we should pause for a few moments and try to picture to ourselves what actually was passing in the minds of scientific men at that middle of the nineteenth century ; that we should try, in a word, to catch the spirit of the time.

Looking back from the vantage-point given us by the passage of seventy years, and standing therefore on an elevation which enables the veriest pigmy of to-day to overlook the head of the tallest giant aforetime, it is not difficult to seize this spirit and to see that this mid-century did in fact coincide with the epoch of a far-reaching change in scientific thought. This change was not, in the main, a change in the ideals or objects of scientific research ; it was shown rather in the direction of a fruitful development of new methods, often opening up entirely fresh vistas and giving access to territories before considered quite inaccessible. The development of spectrum analysis and its application to the heavenly bodies will sufficiently illustrate this point. In general physics, electricity and magnetism, light, thermodynamics, and the laws of energy and its transformations, this decade marked the end of the old experimental school and the rise of the new school of mathematical physicists.   In the early years of the period Faraday was still pouring out the *Experimental Researches in Electricity* papers, which for clearness and charm of style, acuteness of insight, and fertility in experiment will always remain classics, and should be read by every young scientific aspirant whatever branch of science he intends to follow.   The *Philosophical Transactions* for 1851 contain no less than four of these memoirs out of a total of twelve papers.   The last number appeared in the volume for 1856, and in the next year his Bakerian Lecture, *Experimental Relation of Gold and other Metals to Light*, was the last of his great memoirs.   He was then sixty-six, and the remaining ten years of his life were naturally a period of diminishing activity.

Thus ended the stage in which the physicist was compelled to rely mainly upon experiment, and in the next stage experiment tended to become the vehicle of verification rather than of investigation.   At the same time, when Faraday was approaching the end of his labours, William Thomson, then a young man under thirty but with already a continental reputation, was engaged in laying the foundations of thermodynamics and in resolutely clearing away the last difficulties that stood in the way of the full acceptance of the principle of the conservation of energy.   Joule's great memoir, *On the Mechanical Equivalent of Heat*, had been read to the Royal Society in 1849 June, and there only remained to explain the apparent paradox that while a definite amount of heat was exactly equivalent to a definite amount of work, even a perfect engine could, as shown by Carnot's reasoning, only develop a fraction of the total.   What, then, became of the energy apparently lost, and, if lost, where was the conservation of energy ?   The solution was soon apparent to Thomson's acute mind, who saw that it lay in the distinction between the total energy of any system and the available energy, and with this solution the foundation-stone of thermodynamics was laid and the basis of all modern development of energy production firmly fixed.   We may therefore fairly claim that in the year 1851 the two fundamental principles of physical science, principles which neither rearrangements of time and space, nor new conceptions of matter and force have yet shaken, the conservation of energy and the second law of thermodynamics, were defined in terms which would stand to-day and were finally accepted in their present form.   Just about this time another young man of the same school, James Clerk Maxwell, had taken his degree, and thinking of embarking on the study of electricity, was asking advice as to what books on the subject he should read, trying his hand in the meantime on a very difficult problem of astronomical dynamics, the constitution and stability of Saturn's rings.   While in another direction G. G. Stokes was cutting out new paths in the theory of light and taking the place that he held so long of being the ultimate court of appeal on subjects that were sufficiently difficult to baffle all other enquirers. Just by way of a date point, we may note that the *Change of Refrangibility of Light* was read to the Royal Society in two parts, 1852 May and 1853 June respectively.

Turning to astronomy, we find that a change no less fundamental was coming, and though we should, perhaps, not be justified in claiming for the decade under review its actual arrival, we may confidently assert that, looking back now on these ten years, we see in them a period of great development, containing promise of changes even more profound. Anyhow, it is clear that the outlook and aspirations of the astronomer of 1860 were of a different nature from those of his predecessor of 1850. With the invention of photography and the discovery of spectrum analysis, the astronomer's powers were multiplied and the whole scope and possibilities of his science enormously enlarged. Up to 1850 the only photographic process known was the daguerreotype, a method producing pictures of exquisite fineness of detail but demanding high technical skill, and moreover requiring exposures of such length that it was useless as an accessory to the telescope except for the sun. The first recorded astronomical photograph is one of the total solar eclipse of 1851 July 28, for which see volume **41** of the *Memoirs*, Royal Astronomical Society. It shows the corona extending from the limb for about one-fifth of the diameter of the moon. In 1850 the collodion wet plate was invented, and a photographic method with a not too difficult technique and requiring exposures of about one-thirtieth of those previously necessary was thus placed in the hands of experimenters. Sir David Brewster, in his Presidential address to the Edinburgh meeting of the British Association held in that year, devoted a large part of his time to the subject of astronomy, and said, " Though but slightly connected with astronomy, I cannot omit calling your attention to the great improvements, I may call them discoveries, which have been recently made in photography." This view, that the connection between astronomy and photography was slight, was, however, not shared by others, who saw in the new science a most promising addition to the astronomer's tools, and lost no time in attempting to bring it into service.

An early and successful experimenter was Warren de la Rue, an esteemed Fellow of the Society who obtained the Gold Medal in 1862 and became President in 1864. He devoted himself with great energy and at considerable outlay to the construction of a new telescope specially designed for this work, and secured numberless very beautiful photographs of the moon, also some of stars and planets. Another successful exponent of the new art was Mr. Hartnup, of the Liverpool Observatory, who in 1854 obtained pictures of the moon said " to have outstripped all other attempts made elsewhere." He himself was disappointed with his results, a disappointment not unnatural when we realise that his confidence in the sharpness of image and minuteness of detail in his negatives led him to demand that they should stand enlargement from 1·3 inches to 50 feet, equivalent to 460 diameters. A modern dry plate at such an enlargement would show little more than the grain of the silver deposit, and it is no small tribute to the excellence of these early wet plates that any picture at all remained.

In 1854 it was decided by the Royal Society that a photo-heliograph should be established at Kew for the purpose of making a daily record of the state of the solar surface. De la Rue undertook the supervision of the design and erection of the instrument, and it was put into permanent use in 1858 March, remaining at Kew until its transfer to the Royal Observatory, Greenwich, in 1872.

In that same year (1858) the whole astronomical world was thrilled by the appearance of a great comet (Donati's), which for many weeks presented a spectacle of extraordinary beauty and interest. Attempts were made to photograph it, but there was then no instrument provided with the guiding accessories necessary for keeping the image fixed on the plate during a prolonged exposure. De la Rue could obtain nothing in sixty seconds, and the only recorded photograph is one taken by Mr. Usherwood on Walton Common with a stationary camera furnished with a portrait lens of short focus. We can only regret that no possessor of an equatoreal thought of the device of strapping such a camera on to his telescope, which would have given with a short exposure a record of this unique celestial object. We must content ourselves with noting the fact that Mr. Usherwood's was the first photograph taken of a comet.

It would not be right to leave this subject of photography without some allusion, however brief, to the work which was being done at the other side of the Atlantic. Two names are prominent, G. P. Bond at Harvard (1851) and L. M. Rutherfurd from 1864. In both cases their main object was the same, to develop photo-graphy as a method of precision for the delineation and recording of star positions ; to what splendid development this has attained at Harvard is well known to all and need not be described here.

Concurrently with this advance in celestial photography, a great improvement was made in the reflecting telescope by the substitution of the silver-on-glass for the speculum metal mirror. The previous decade had seen the completion of Lord Rosse's 6-foot telescope and its application to the resolution of nebulæ, a task in which it proved so efficient that it tended to mislead astronomers into the incautious assumption that, because many nebulæ hitherto supposed irresolvable, were supposed to be seen in the field of this powerful instrument separated into their component stars, given sufficient magnification all nebulæ would thus prove to be distant clusters.   Lord Rosse was awarded one of the Royal Medals by the Royal Society in 1851, but he never received our Gold Medal, a somewhat curious omission which it is difficult to justify, though the explanation is presumably to be found in the fact that he was pre-eminent not primarily as an observer but rather as an engineer, who developed to a high level of perfection the art of casting, grinding, figuring, and polishing large metal specula and in mounting them for practical observing.

Another name which will always be associated with the early use of large reflecting telescopes is that of William Lassell, who had been awarded the Gold Medal in 1849 and served in the office of President in 1870–72.   Lassell was both a constructor and a skilful observer, and not satisfied with our murky skies, he transported his instrument, a reflector of 2-feet aperture and 20-feet focal length, to Malta in 1852, and observed with it there for some time, replacing it later with an instrument of double the aperture, which he used at Malta in 1862–65.   The invention of the silver-on-glass reflector threw the metal mirror out of commission, and a few years ago it might have quite confidently been maintained that no large metal speculum would ever again be undertaken.   Lately, when we have seen the almost insuperable difficulties of casting very large glass discs (witness the case of the Mount Wilson 100-inch, where it took years before a suitable one could be obtained), it is not quite so certain that we or our descendants may not see a reversion to metal.   The casting of a metal disc of any desired size presents no insurmountable engineering problems.

The last year of our decade saw the most momentous and far-reaching enlargement of the boundaries of astronomy : the application by Bunsen and Kirchhoff of the principle of spectrum analysis (found by Stokes in 1853) to the determination of the constituents of the solar atmosphere.   Kirchhoff's first paper on this subject was read before the Berlin Academy on 1859 December 15.   His full memoir did not appear till eighteen months later, and the early astronomical applications of the new research thus lie within the next decade.   We may, however, just note that this announcement, so important to us and marking off so clearly in our estimation the opening of a new epoch, does not seem to have excited very much attention among astronomers at the time.   The Astronomer Royal in his suggestions for the observation of the solar eclipse of July 1860, did not allude to the possibility of work of this nature ; nor does the word spectrum appear in the index to the *Monthly Notices* until 1863, when Airy, at the March meeting, described a spectroscope used at Greenwich on fixed stars. The number for 1856 May had, however, contained a short "Description of an Observatory erected at Upper Tulse Hill," by William Huggins, who was then devoting himself to the study of astronomy with the intention of making it his life work. How quickly he seized the splendid promise of Kirchhoff's work, and to what fruitful use he put it, is familiar to all, but, in any case, falls outside our period and cannot be told here.

The year 1850 found astronomers still greatly interested in the discovery of new minor planets; ten of these bodies were then known, and their number was being added to yearly. In 1853 February, twenty-three were known, and by 1860 January, the total had risen to fifty-seven. While the number known was small, it was confidently expected that the end would soon be reached and the whole group named and accounted for. Thus we find the Report of the Council for the year ending 1851 February, after recording the discovery of three more, said : " A rate of increase among the known members of the solar system which can hardly be expected to continue very long." When, however, it began to be suspected that the rate of discovery was a function of the increase of telescopic power, that greater efficiency and higher magnification simply meant that the existence of smaller and smaller planets was disclosed and that their total number was, for all practical purposes, infinite, the interest of the search waned. Now the only justification that remains for continuing this search is the possibility of finding one with an exceptional orbit, such as was the case with Eros, where the fact that its orbit at times comes inside that of Mars makes it the most accurate gauge for the measurement of the solar parallax.

In one particular direction the Astronomical Society found the rapid multiplication of these bodies a decided embarrassment. Their ephemerides had always been published in the *Monthly Notices*, but the bulk of this matter became too great, and in 1854 February the Editor reported that, as communications of this nature could be published with greater regularity as well as earlier and more fully in the *Astronomische Nachrichten*, they would disappear from the *Monthly Notices*.

The two most active observers and discoverers of these little bodies at this time were J. R. Hind, and A. de Gasparis of Naples. Both received the Gold Medal, Gasparis in 1851 and Hind in 1853. The President in the first-named year was Airy, and in his address he set forth the three known ways of detection, viz., accident, physical properties, *e.g.* recognition of the planet by its disc, and careful comparison with star charts, and he lamented that this country possessed no charts at all comparable with the Berlin series, a defect so markedly brought to light in the history of the discovery of Neptune. Hind worked at the observatory established by George Bishop, who was for many years Treasurer of the Society, in the grounds of his house, South Villa, Regent's Park, on the site now occupied by Bedford College. He at first used the Berlin charts, but finding that they were deficient of the fainter stars, nothing below the tenth magnitude being shown, and that they did not include the whole of the region about the ecliptic, he set himself to make another set of charts containing all stars down to the eleventh magnitude within 3° of the ecliptic, and with the aid of these was able to make further discoveries. This total was ten planets,—eight up to the time when he received the Medal, and two shortly afterwards; but as he was then appointed Superintendent of the *Nautical Almanac*, he was compelled to discontinue active observing work. With the exception of three found by Pogson at Oxford during 1856 and 1857, no further minor planets can be credited to any observer in this country.

In other directions there was great activity in these years in the construction of star charts and catalogues: 1850 was the epoch of the British Association Catalogue, due to the indefatigable labours of our energetic Fellow, Francis Baily, which recorded every star down to the sixth magnitude of which reliable observations could be found; 1855 must always remain a memorable date to astronomers, as being the epoch of the *Bonn Durchmusterung*, generally called the B.D., which far excelled all previous star atlases. Originally carried out by Argelander from the N. Pole to 2° beyond the Equator, and continued by his successor Schönfeld down to 23° south of the Equator, it showed every star down to magnitude $9\frac{1}{2}$, and was for many years, in fact until, to some extent, superseded by modern photographic charts, invaluable for comet searching and similar tasks. The intensive scrutiny for minor planets on the Continent gave rise to the production of beautiful ecliptic charts by Chacornac and the brothers Henry.

There was one astronomical advance of prime importance made about this time, to which it is not easy to fix a precise date, Schwabe's proof of the periodicity of sun-spots. He, it is true, announced in 1844 that he had got indications of a ten-year period, but the announcement excited no attention and apparently met with little belief. In 1850, Humboldt, in the third volume of his *Kosmos*, gave a table of the sun-spot statistics as observed by Schwabe from 1826 to date, in which the periodicity was so evident that it could not be overlooked. Schwabe was awarded the Gold Medal in 1857, the President being M. J. Johnson, the Radcliffe

Observer, who in an excellent address paid full tribute to the medallist's unexampled perseverance and industry. In 1852, R. Wolf announced a corrected period of 11·11 years, and in the same year the connection between sun-spots and the magnetic elements on the earth was shown independently by three men : Gen. Sabine in England, Wolf at Berne, and A. Gautier at Geneva.

We must also claim for this decade the actual completion and publication of a gigantic work, not perhaps strictly astronomical, but with an intimate connection with astronomy, the Principal Triangulation of the United Kingdom. The scientific direction and control of the calculations were in the hands of A. R. Clarke, who was elected a Fellow on 1850 March 8, and continued in the Society until his death a few years ago. He was never awarded the Gold Medal, an omission which some of the geodesists in the Society regret, but they admit at the same time that there is no instance of an award for geodetic work pure and simple. Clarke's *Geodesy* still remains the best English book on the science, though the great activity on the Continent in recent years has produced publications which now leave it, in certain directions, somewhat out of date.

In 1852 there occurred the death of one of the founders of the Ordnance Survey, Thomas Colby, who was concerned with the triangulation of England from its beginning in the early years of the century and took a leading part in it up to the year 1847. His obituary notice covers nearly fourteen pages of the Report of the Council to the Annual Meeting on 1853 February 11. Among many other activities, he was the designer of the compensation bars used in the measurement of the Salisbury Plain and other bases.

The publication and discussion of these geodetic results drew renewed attention to an old problem, the determination of the mean density of the earth. Three methods were available for its solution. Firstly, the comparison of the earth's attraction with the horizontal attraction exercised by a mountain mass. This method had been used by the Ordnance Survey, and an account of it was presented by Col. James, the Director-General, to the Royal Society in 1856 February. A second possible method was a comparison of the attraction at the surface with that found at a definite distance below the surface. This was tried by the Astronomer Royal at Harton Colliery in 1854, with every precaution that his skill and experience could suggest. The actual result was disappointing, the value found, 6·566, being now known to be very considerably in excess of the truth. The third method, and the only one of these three which would now be reckoned of any value, is the Cavendish experiment, wherein the earth's attraction is directly compared, by means of a torsion balance, which can be made of almost any required degree of sensitiveness, with the attraction

of a small mass of metal of known weight and density.   This experiment had been repeated by Francis Baily, who had obtained a
value of 5·67; while the Ordnance Survey, by observing the deflection of the plumb line at Arthur's Seat, got 5·316.   Cornu and
Baille found by the torsion balance 5·56; while Boys, in his skilful
and laborious repetition of the same experiment (1895), using
quartz fibres and every possible refinement, got 5·527.

On 1855 August 7 there died one of the most loved and respected
Fellows of the Society, the Rev. R. Sheepshanks, of whom the
Astronomer Royal said, " a man whose equal in talent and perseverance, in disinterestedness, in love of justice and truth I have scarcely
known."   He died as the direct result of overwork on a laborious
task on which he had been engaged for more than eleven years,
the construction of the new standard yard, involving tens
of thousands of measurements and comparisons with standards
of different lengths, different materials and varying methods of
marking the fiduciary points.   This work had been determined upon
by a Royal Commission in 1843 and had then been entrusted to
Baily.   He, however, died in 1844, and Sheepshanks took it up and
gave unremitting attention to it till his death.

Sheepshanks had an extraordinarily skilful eye with the
micrometer, and it was stated that his comparisons were " so far
superior to those of all preceding experimenters, including Kater
and Baily, as to defy all competition on the ground of accuracy."
He left the work so far completed that no further measurements
were required, and it was subsequently put in form and published
by Airy.   He was a generous benefactor to the Society and to the
University of Cambridge.   Eighteen months after his death his
sister, carrying out his wishes, though these were not actually
embodied in his will, presented all his astronomical and other
scientific instruments, now forming forty-three items in the list
of instruments in our possession.   With all his energy and capacity
he was a very modest man, and always refused to be nominated
for the office of President, though it was an office for which he was
pre-eminently fitted and to which he would have received the
warmest welcome both from the Council and from the Fellows.
The acceptance of it was, in fact, often pressed upon him.

The capture and occupation of Lucknow brought to the notice
of astronomers the history and ultimate fate of one of the most
short-lived observatories ever established, the one founded by the
King of Oude in 1841 and abolished by him, when the new toy had
ceased to amuse, in 1848.   The observatory was well equipped and
furnished with instruments of the first order by Troughton and
Simms.   A Fellow of the Society, Lt.-Col. Wilcox, had been
appointed Director, and the new institution began work under the

THE REV. RICHARD SHEEPSHANKS
(1794–1855)

To face p. 118.]

happiest auspices ; the Director was himself an energetic observer. The native subordinates were easily trained and most trustworthy, as indeed the experience of the Indian Survey Department had consistently proved ; the climate was good, and there was no other observatory at the same low latitude so well furnished with instruments of power and precision. When, however, the King discovered that it was not enough to build and equip an observatory and pay an observer, but that it was also necessary to publish the results and that this publication called for an annually recurring expenditure, his interest rapidly waned, and he finally, in 1848, discharged the Director and placed the papers and instruments in the charge of a native officer who knew neither English nor astronomy. After the Mutiny, when the city was entered by the British forces, there happened to be one Fellow of the Society present with the troops, Lieut. J. F. Tennant, then a young officer on the Survey, who ultimately rose to high rank and was President in 1890–92. He was at special pains to ascertain what had happened to the observatory, and it appeared that while the structure was more or less intact, all the instruments, in fact everything of metal, had been removed, and that it was beyond hope that anything in the shape of an instrument could be recovered. The records had long since been eaten by white ants. Thus perished the Lucknow Observatory.

Allusion has already been made in connection with the subject of astronomical photography to the solar eclipse of 1851 July 28. This eclipse was visible in Scandinavia at many easily accessible points, and a very large number of Fellows took advantage of the opportunity. It may fairly be said to have been the first eclipse of which extensive co-ordinated observations were attempted. The observations were necessarily confined to drawings of the corona and prominences, times of contacts, and notes of the effect upon men, animals, and birds. The Astronomer Royal was at Gottenborg in Sweden, and among others were Hind ; Dawes, a most industrious and painstaking observer, joint re-discoverer with Bond of Saturn's dusky or crape ring; Carrington, then in charge of the University Observatory at Durham, subsequently to be famous for his long-continued series of observations of sun-spots, leading him to the discovery of their drift, or variation in apparent rotation period with the solar latitude ; Piazzi Smyth, the brilliant Astronomer Royal for Scotland; Lassell; and Dunkin of Greenwich. The eclipse was a fine one, with a remarkable hook-shaped prominence and a bright corona visible to a distance from the limb equal to about half the moon's diameter. The usual effects upon the animal world were noted, and the behaviour of the human spectators varied from an old lady who lit a candle to continue her work, to fishermen who showed great terror, and to the inhabitants of one village who donned their best clothes in honour of the event.

The name of Piazzi Smyth will recall an enterprise which he undertook shortly after : the transport of a telescope to the Peak of Teneriffe, and the conduct of regular observations there for many months in order to ascertain exactly what the astronomer stood to gain by elevating himself some thousands of feet. It was not for many years after this that a permanent mountain observatory was established ; but Piazzi Smyth's expedition, described as it was with all the picturesque wealth of language of which he was so well capable, must always remain memorable as the first real attempt of an astronomer to free himself from the restrictions imposed by the opacity or turbidity of our atmosphere. On the other hand, the fact, already noted, that Hind, observing in London, discovered no less than ten new minor planets would seem to indicate that even our foggy skies are not quite so antagonistic as might have been supposed to observations of the utmost delicacy.

This summary, necessarily very short, and indeed almost fragmentary, will give some indication of the major astronomical activities of the time, especially those with which our Fellows were concerned, for, as has doubtless been fully appreciated by the reader, we have paid little attention to work done or discoveries made abroad. The time was clearly one of great progress, while, as already insisted upon, it was essentially a time of transition, leading from the old astronomy of observation, the construction of star charts and catalogues, the detection of minor and the close scrutiny of the surfaces of the major planets, to physical astronomy in its new meaning inaugurated in the next decade. The general interest in astronomy, both among scientific men engaged in other lines of work and among the educated public, was very great, possibly much greater than it is at the present time. A good gauge of this is the content of the Presidential addresses delivered at the opening of the annual meetings of the British Association. In 1850, Sir D. Brewster was President, and devoted a large part of his address to our subject. He naturally dealt with the revelations of Lord Rosse's telescope, and alluded to the recent discoveries by Lassell of the satellite of Neptune and the eighth satellite of Saturn ; also to the eleven then known minor planets. The next year he was replaced by Airy, who, of course, devoted his whole address to his own beloved science. He claimed that the progress of astronomy in the past year had been very great, and among a multitude of other points dealt with the new Foucault pendulum experiment. The Belfast meeting of 1852 found Sabine President, the discoverer of the correlation between sun-spot periodicity and magnetic elements ; and in the next year the foundation of the American *Ephemeris and Nautical Almanac*, the first year of issue being for 1855, was welcomed.

At Liverpool in 1854 the Earl of Harrowby obtained assistance for the astronomical section of the address, evidently considered a necessary part of it, and included a report prepared by Challis on the present state of the science.   Again in 1857 at Dublin the Rev. Humphrey Lloyd began with astronomy, and gave a competent summary of recent progress.

One literary event during our decade must not pass unnoticed, the publication of Grant's *History of Physical Astronomy*.   For this work the author received the Gold Medal in 1856, an honour never before accorded for literary service.   The book is well known to all astronomers and will long continue to be read.

The internal or domestic history of the Society during these ten years was one of quiet activity and progress, and was marked by none of those clashes of personalities which, while often unpleasant to contemporary spectators, furnish the memorialist with his most telling paragraphs.   The annual meeting of 1850 February elected Airy as President for his fourth year.   The Treasurer was Bishop, the secretaries A. De Morgan, a most devoted friend of the Society, who served in that office for sixteen years (1831–39 and 1847–55), and Capt. Manners ; the Foreign Secretary, Hind ; and among the Council were Adams, James Glaisher, the famous meteorologist and balloonist, and John Lee, a generous benefactor of the Society. Among the Vice-Presidents were Main, who subsequently became Radcliffe Observer, and Sheepshanks.   The next year Adams replaced Airy, the other officers and members of Council remaining for the most part unchanged.

In this year the Council were much alarmed at a proposal of the Government to erect offices for the scientific societies on ground held by the Commissioners of the 1851 exhibition at Kensington Gore. They were unanimous in denouncing this scheme, and expressed themselves as more than satisfied with the rooms they had occupied in Somerset House for eighteen years.   They maintained that the removal of the Society to a " distant suburb " would compel the resignation of many of the working Fellows, and they said that if the scheme were proceeded with they would have to petition Her Majesty for leave to remain in their present apartments ; should this not be granted, they would prefer to hire their own quarters rather than exile themselves.   The Royal Society Council was equally unanimous in rejecting the Government project, and nothing more was heard of it.   In 1854 the question of removal again came up, when the rebuilding of Burlington House as a home for the Royal Academy and for all the leading scientific societies became a practical possibility.   The Society was, however, quite comfortable

at Somerset House, and the Council at first "unanimously de-
precated" any removal; if, however, removal was insisted upon,
they formulated certain conditions. It must be remembered that
the original Government scheme for Burlington House would have
given accommodation of a different character from that now provided. The intention was to give all the societies offices of sufficient
size for their routine clerical work and rooms for Councils or Committees to meet, while the rooms for scientific or general meetings
were to be common to all the societies, and used by each in turn
in accord with a mutually arranged calendar. The libraries of
all the societies were also to be amalgamated into one, to which all
Fellows and Members would have access. In many ways the
original proposal was a thoroughly businesslike and practical one,
and would have resulted in great saving of space, while the effective
accommodation for each science would have been quite as ample as
at present. The individuality of the various societies was, however,
too strong, the measure of co-operation proposed proved impossible,
and the plan was ultimately abandoned. Several of the societies,
our own included, were prepared to accept a common meeting-
room, but they could not reconcile themselves to a common library.
The other conditions laid down by the Council were that they
should have not less space than they had at Somerset House, that
quarters should be provided for a resident Assistant Secretary, and
that they should be put to no expense. At the same time, the
officers were not anticipating an early move. In 1854 July, De
Morgan wrote to Admiral Smyth on this subject : " All is going on
well as to the Government proceedings. We shall not be stirred
these ten years, I augur. You know the story of the birds in the
nest listening to the farmer plotting how to cut the corn. Now
Government is a man who cannot work for himself. He works
through people who *report*. Deep calleth unto deep—that is, one
office reports to another, and the other refers back, and then they
consider, and red tape becomes grey before they have settled how
to proceed. And if you give them six months' start and set a snail
at them, the snail beats them by a thousand lengths ; and then
there is a change of Ministry and a new report to ' my lords,'
and ' my lords ' make a minute, which means in time a year, and
so *ad infinitum*."

De Morgan over-estimated the speed of a Government Department ; it was twenty years before the shift of quarters was accomplished.

In 1853, Airy again took the Presidential chair, and among new
members of Council were De la Rue, then just beginning his work
in astronomical photography, and Grant, whose *History* has already
been mentioned. The two years of Airy's Presidentship were in

the main uneventful. The Medal in 1854 was awarded to Charles Rümker for his astronomical observations in general, and for his *Catalogue* of twelve thousand stars in particular. In 1855 it was given to Dawes for his long-continued devotion to astronomy, his numerous contributions to the science and the excellence of his observations. The Presidential address on that occasion was a model of terseness and brevity and filled barely two pages of the *Monthly Notices*. All the addresses of this period were in fact much shorter than we have been accustomed to recently, and though it is doubtless often of great value to have a reasoned review of a man's work put before us, and such review, if it is to be complete, cannot be very short, it is nevertheless possible that the present generation might learn something of the art of condensation by studying the models set by their great predecessors.

In 1855 the choice of the Council fell upon M. J. Johnson, the Radcliffe observer, and De Morgan yielded the secretaryship, which he had held so long, to De la Rue. Carrington from Durham, and W. Simms, of the famous firm of Troughton and Simms, took their seats at the Council table. In 1856 January the Council, which had become somewhat uneasy at the large number of Associates elected, appointed a Committee to consider the question and recommend whether any limitation should be imposed. In July the Committee made a long and careful report. It appeared that the number had risen from about 21, at which it stood in the early years of the Society's existence, to 56. The names were carefully scrutinised and it was found that no one had been appointed who did not reflect honour upon the Society and that, in general, the selection could not have been better done. The Committee, however, thought it would be desirable in future to have an understanding, not explicitly embodied in a Bye-law, that the number should be restricted to about 50. This was accepted by the Council and has remained a working principle since that date. Even with the enlarged astronomical developments of the present times the number appears to be sufficient to include everyone of real distinction, and it would be hard to find any instance of a foreign astronomer of high merit who has not been included in the list.

It may be noted that the original idea of the qualifications and functions of an Associate differed in one important particular from our present conceptions. Now we look upon the election as an honour to a foreign astronomer, accorded him upon the sole basis of his services to astronomy. In the early years of the Society, and still persisting in 1856, it was held that an Associate should be recommended for election not only in recognition of his past achievements but also in hopes of his future services, and great stress was laid on the importance of selecting such men as could

" usefully co-operate with the Society and assist it in carrying out schemes requiring organisation and division of labour." This consideration somewhat modified the type of man recommended. Thus while the Committee were of opinion that the list should contain " Directors of foreign observatories of deserved reputation and of such foreign Professors of Astronomy as have conspicuously added to the science by theoretical investigations," they were very chary in admitting any claims on the part of those who did not hold official positions. Such were considered as having little or no value for these co-ordinated researches. The amateur astronomer the Committee would have entirely excluded, and they specifically recommended " that, though subordinate foreign astronomers may occasionally in cases of extraordinary merit be usefully put on the list of members, yet that want of an independent position must be considered *per se* as a disqualification only to be set aside by some remarkable discovery or some elaborate astronomical work of acknowledged merit." Note particularly the word " usefully " as indicating the bias of the Committee.

The hope that Associates could be made use of never appears to have been practically acted upon, and has long since been abandoned. In fact, it is now recognised that the Society is not a suitable body for carrying out schemes requiring organisation and division of labour.

The Council of those times was troubled with one task lying somewhat outside their normal functions, the exercise of a piece of ecclesiastical patronage arising out of the gift to the Society by John Lee of the advowsons of the livings of Hartwell and Stone. Thus we find in 1855 the Council had to present a new incumbent to the living of Hartwell-cum-Hampden-Parva, and again in 1859 had to fill a similar vacancy at Stone. In the first case there were six applicants, and the selection was made easier by the fact that the duty had for some years been performed by a clergyman residing in the place. The gentleman, the Rev. C. Lowndes, was himself an amateur astronomer and had, aided by the generosity of Mr. Lee, built himself a small observatory in the Rectory grounds. The Council were therefore in the fortunate position of being able to satisfy both doctrinal and scientific needs, to appoint a Rector who was already well known and respected in the locality, and to help towards the continuation of work in the observatory. On other occasions there was no such clean-cut issue before them, and in filling up the living of Stone they were apparently guided only by testimonials. Ultimately, these two advowsons were repurchased from the Society by Mr. Lee's heir in 1879, and the Council was relieved of a distasteful duty for which it had no qualifications.

Towards the end of 1856 another matter of procedure was remodelled, the method of voting for the award of the Gold Medal was modified and the present system established. Up to that date the practice had been to receive nominations or proposals at the November meeting, to take no action in December and to select the recipient, from among those proposed, in January. The Council meeting of January therefore had all the nominations before it, and was directed by the Bye-law to " decide to which of the persons so proposed the Medal shall be given at the ensuing Annual General Meeting." No explicit power was placed in the hands of the Council to vote that a Medal should not be awarded in any given year, but as the award required a three-fourth majority it would obviously occasionally occur that the candidate obtaining the suffrage of the majority could not satisfy three-fourths of the Council, and no election would be made. Up to 1850 no Medal had been awarded on six occasions. This procedure was found to have certain admitted difficulties, and it was decided to change it by adoption of the principle of dividing the election into two steps ; a selection from among the candidates in December and the election of the selected candidate in January by a three-fourths majority. It should be particularly noted that the question on which they are asked to vote in December is different from that on which the January vote is taken. In December the vote taken is as to the relative merits of the candidates ; in January, as to absolute merit, *i.e.* whether the astronomical work of the selected candidate is up to the standard set for the Medal. The January voting is not, as has sometimes been assumed, a confirmation of the previous selection, but in the words of the Bye-law is a vote " whether the nominee then before the Council shall or shall not receive the Gold Medal," *i.e.* is his work of such merit as to deserve unquestionably this high honour ? There is therefore no inconsistency in a member of Council who votes for a particular candidate in December and against him in January ; he is merely expressing his view that while the nominee is the best of those proposed, he is not quite of sufficient distinction for the award.

This principle was clearly laid down by the Committee of 1856, and the Bye-laws were then redrafted in the form which, with one minor amendment, now stands. In 1857, Bishop was elected as President, and S. C. Whitbread took his place as Treasurer. Whitbread was known rather as a meteorologist than an astronomer, having been one of three original founders of the Meteorological Society, with Lee and James Glaisher, in 1850. He was a most efficient Treasurer, and held the office for twenty-one years. He was an absolute terror to defaulters in arrear with their contributions, and used to visit them personally and ask them to explain their conduct before he recommended the Council to expel them.

In fact, the whole process of removing a Fellow's name from the list on account of failure to pay his dues was then a much more formidable affair than at present.   Failing to get any satisfaction at a personal visit, the Treasurer was obliged to recommend expulsion, a ceremony carried out with the utmost publicity at a Special General Meeting called for the express purpose.   Nowadays the Council is more charitable, and recognising that a failure to pay may be due more to misfortune than to malice, allows the name to disappear quietly from the roll without publicity.

The remaining years of our decade call for little notice.   Bishop served the usual two years as President without ever once taking the chair owing to ill-health, and was replaced in 1859 by the Rev. Robert Main, who in the following year succeeded Johnson as Radcliffe Observer.   Among new members of Council we may note A. Cayley, the famous algebraist, and A. R. Clarke, the geodesist. Admiral Smyth, the author of *A Cycle of Celestial Objects*, returned once more to the Council which he had served so well in previous years.

During the whole period the two publications of the Society, the *Memoirs* and the *Monthly Notices*, grew in size and importance, and may justly be said to have contained almost everything of any permanent value in astronomy that was published in Great Britain.   An old dispute, even in recent years not quite dead, as to the relative position as regards publications of scientific papers between the Royal Society on the one hand and the specialised Societies on the other, arose somewhat acutely at this time. The story of Sir Joseph Banks and his jealousy at the founding of the Astronomical and other societies has already been told in an earlier section of this history.   Long after his time it was, however, still held by many claimants on behalf of the premier society that they had an absolute right to the publication of all scientific memoirs of the first order of importance, and that the others could only claim either work of second-rate merit or, if they cared to do so, might produce abstracts of work already issued by the Royal Society.   It need hardly be pointed out that no question of claim or right arises.   Anybody is entitled to send his papers to any Society of which he is a Fellow, failing that, he must get a Fellow to present it on his behalf, and the choice as to which Society he selects rests exclusively with him.   It has never been seriously proposed, though we do not doubt that many of the out-and-out upholders of the extreme claims of the Royal Society would have supported it, that the Council of a Society such as the Astronomical should, if they judge a paper to be of sufficient merit, pass it on to the Royal Society for publication.   No upper limit has ever been set, or could conceivably ever have been set, to the quality of our *Memoirs*. In default of this it is not easy to see how the Royal Society could hope to enforce a claim to receive the best. Such claim was, however, definitely made, and made, moreover, by one who was both an astronomer and an ex-President of the Royal Society. In 1855, Lord Rosse presented a confidential memorandum to the Council on the expediency of enlarging their number. He said, " In a Council so small it is impossible to secure representation of the leading scientific societies, and it is scarcely to be expected that under such circumstances they will continue to publish inferior papers while they send the best to our *Transactions*." In case it should be suspected that we are here guilty of a breach of official confidence, we hasten to explain that the above extract is public property, having been printed by De Morgan in the *Budget of Paradoxes* nearly two generations ago.

It is, as De Morgan pointed out, not quite easy to see what Lord Rosse meant when he spoke of the societies sending their best to the Royal Society, but the nature of his pretension is abundantly clear. Such a claim was, however, not supported by other astronomers. The *Philosophical Transactious* for 1850–60 contain only four papers of an exclusively astronomical character, and it is quite certain that there was no general acquiescence in the idea that our Society should get only the second best. Lord Rosse himself, though he had been a Fellow from a date within a few years of its foundation, was not one of our ardent supporters. He only once served on the Council (in 1827), and, with one minor exception (in *M.N.*, **14**), never communicated any of his scientific memoirs to us for publication.

A modus vivendi with the Royal Society has long since been tacitly agreed upon, and such a difference can never arise again. In fact now, owing to the great increase in the cost of printing, the question has become inverted, and in place of any jealousy between societies as to what they are asked to publish, they are only too glad to find other bodies who will in any part relieve them of an expensive duty. In 1850–60 the *Monthly Notices* cost about £120 per annum, now they cost nearly £1000. Thus we leave the Society full of energy and enthusiasm, eager for progress and alert for new knowledge. We now, looking back, can see that a splendid day was dawning, though the light was doubtless yet faint and carried hope only to those blessed with the keenest vision. New ideas were arising in every direction. Charles Darwin's *Origin of Species* had just been published, and though to astronomers the notion of evolution, of slow change and development through countless ages, was nothing new, and was to them an accepted factor in the history of the inorganic world, its courageous application to the organic world justified and supported their own speculations and carried with it the hope of further advances. The essential unity of nature, a unity assumed in the earlier nebular hypothesis, and now actually verified for our own system by the demonstration that the sun contained the same elements as the earth, was thus in part proved, while the complete proof, though difficult, was seen not to be impossible. The extension of this same unity to the stellar universe soon followed, and with this extension the ancient science of Astronomy entered upon a richer and a fuller life.

CHAPTER V

THE DECADE 1860–1870. (By H. F. NEWALL)

A DECADE which was so full of activity and achievement in all branches of astronomical progress as that between 1860 and 1870, makes great demands on the self-restraint of an astronomer who is called upon to set forth the history of our Society at that time. There are great temptations to the historian to digress from the strict lines within which he should confine his efforts, and to allow himself to be guided in his references, not only to events in other decades, but also to researches which in truth belong to the history of Astronomy, and not to the history of the Society.

This particular decade, 1860 to 1870, would certainly afford a very interesting chapter in the history of Astronomy. But that is not our present task; still, some indication must be given of the activity of the decade with which we have to deal.

In it we see the application of photographical methods to furnishing the best basis for lunar topography and to recording the complex phenomena of solar eclipses.

We see the development of spectroscopy, not only as affording evidence of the widespread distribution of terrestrial chemical elements throughout the universe, but also as giving proof of the radical distinction between gaseous nebulæ and unresolved star-clusters.

We see the bold and pertinacious attack on the measurement of the line-of-sight velocities of stars by means of the spectroscope.

We see also another triumph of the spectroscope in the discovery of the nature of the solar prominences as outbursts of incandescent gas, and the almost simultaneous discovery of a method of daily observation of such prominences, which hitherto had been disclosed only during total eclipses of the sun.

We see visual methods in the study of the positions and motions of sun-spots replaced by photographic records; but not before the peculiarities of those motions and of the rotation of the sun had been demonstrated.

We see greatly increased activity in the observation of meteors and meteor radiants, and also the establishment of the identity of orbits of certain comets and meteors.

We see great advances in the determination of the solar parallax, great advances in planetary theory, and great advances in lunar theory.

We see the development of the idea that the lunar acceleration must be connected with tidal friction and the consequent lengthening of the day.

We see the figure of the earth defined in terms of a possible ellipsoid.

We see the number of minor planets carried well beyond the first century, the hundredth being named after that goddess of mischief, Hecate.

We see the completion of Argelander's *Bonn Durchmusterung*, and the publication of the following notable contributions :—

> Sir John Herschel's *General Catalogue of* 5079 *Nebulæ* (1864).
> Lassell's (Marth's) *Catalogue of* 600 *New Nebulæ* (1867).
> Rosse's *Observations of Nebulæ* with his 6-foot Speculum (1861).
> Carrington's Memoir on Sun Spots (1863).
> De la Rue's Memoir on the Solar Eclipse, 1860 July 18 (1862).
> Tennant's Memoir on the Solar Eclipse, 1868 August 17–18 (1869).
> De la Rue and Balfour Stewart's *Heliographic Positions and Areas of Sun Spots* (1862).
> Huggins and Miller's work on The Spectra of the Stars and Nebulæ (1864).
> Dunkin's Memoir on *the Motion of the Sun in Space* (1864).
> Lockyer's early work in *Solar Physics* (1868).
> Huggins's Memoir on his Attempt to determine whether Stars are moving towards or from the Earth (1868).
> Bond's monograph on Donati's Comet (1862).
> Thomson's paper on *The Rigidity of the Earth* (1863).
> Airy's work on *The Diurnal Variation of Magnetic Elements*.
> Chauvenet's *Spherical and Practical Astronomy* (1863).
> Watson's *Theoretical Astronomy* (1868).

The decade was notable, too, for several remarkable astronomical events.

The earth passed through the tail of a comet (1861).

The star T Coronæ blazed out with great brilliance (1866 May).

The long-expected November meteors made their memorable display (1866).

There were two solar eclipses which have become historically notable : 1860 July 18, and 1868 August 17–18.

This decade saw the completion of several large telescopes :

Lassell's 4-foot speculum reflector (1861), which Lassell and Marth used in observations of satellites and nebulæ at Malta, 1861. It was broken up before Lassell's death in 1880.

The Melbourne 4-foot speculum Cassegrain reflector (1869), built by T. Grubb, and used by Le Sueur and Ellery in observations of southern nebulæ, very little of which has been published.

The Dearborn 18½-inch refractor, 1864, built by Alvan Clark & Sons, and used by Burnham in some of his observations of double stars.

The Newall 25-inch refractor, 1869–70, built by T. Cooke & Sons, and now at Cambridge.

The Directors of Observatories in this mid-century decade may be recalled as follows :—

Greenwich	.	.	G. B. Airy	.	.	.	1835–1881	†1892
Cape	.	.	Sir T. Maclear .	.	.	.	1833–1870	†1876
Edinburgh	.	.	Piazzi Smyth .	.	.	.	1845–1888	†1900
Dunsink	.	.	Sir W. R. Hamilton .	.	.	.	1827–1865	†1865
			F. F. E. Brünnow	.	.	.	1865–1874	†1891
Cambridge	.	.	J. C. Adams	.	.	.	1861–1892	†1892
Oxford (Radcliffe)	.	.	Rev. R. Main .	.	.	.	1860–1878	†1878
Durham	.	.	Rev. Temple Chevalier	.	.	.	1842–1873	†1873
Liverpool .	.	.	John Hartnup	.	.	.	1843–1885	†1885
Glasgow	.	.	Rob. Grant	.	.	.	1859–1892	†1892
Armagh	.	.	T. R. Robinson	.	.	.	1823–1882	†1882
Madras	.	.	N. R. Pogson .	.	.	.	1860–1891	†1891
Melbourne	.	.	R. L. J. Ellery	.	.	.	1863–1895	†1908
Nautical Almanac	.	.	J. R. Hind	.	.	.	1853–1891	†1895
Pulkowa .	.	.	Otto Struve	.	.	.	1862–1889	†1905
Paris	.	.	Le Verrier	.	.	.	1853–1870	†1877
Berlin	.	.	J. F. Encke	.	.	.	1825–1864	†1865
			W. Foerster	.	.	.	1865–1904	†1921
Göttingen	.	.	Klinkerfues	.	.	.	1859–1884	†1884
Bonn	.	.	Argelander	.	.	.	1837–1875	†1875
Copenhagen	.	.	D'Arrest	.	.	.	1857–1875	†1875
Athens	.	.	Schmidt	.	.	.	1858–1884	†1884

Among the references in the Annual Reports of the Council to private observatories, we find the following names : Carrington, Lassell, De la Rue, Lee (Hartwell), Selwyn, Dawes, Huggins, Wrottesley, Rosse, Lockyer, Webb, Howlett.

As we look through the records from the time of its first inception, the Society seems to have flourished most especially by reason of the confidence which it showed that the science which it was formed to encourage, could best be fostered by giving complete freedom to its members to prosecute researches chosen by them without any official pressure in special directions.

The absence of imposed guidance and the freedom of its individual members is a feature which cannot fail to strike us now, living in these days of tendency towards organisation of scientific work.

Fortunately our object is not to formulate new schemes, but rather to recognise the tradition handed down to us and do homage to the men of old. We can no more anticipate how our science will stand a hundred years hence than our founders could predict whither their methods and their labours would lead us.

To an astronomer who looks back through the century, the feature which will probably strike him most is rather how little in essential details is the change of outlook on the problems to be studied and solved. The work of Newton had laid down lines along which men's minds had contentedly followed ; and every new observation that was made seemed to fall naturally into its place in the system of the universe which Newton had formulated. Even the most recent developments of Einstein and his followers may be regarded, not so much as upsetting the Newtonian universe, but rather as affording an opportunity of gauging phenomena that present themselves in conditions which transcend those contemplated in Newton's philosophy. The stimulating influence of the Society on the production of the work can hardly be over-estimated. The Society was a focus, which performed a double service of the greatest value. It served to bring the professional workers in contact, not only with one another, but also with the large body of amateur astronomers, who in this country have always formed so marked a proportion of the whole constituency. The meetings afforded an interested audience, before whom investigators were proud to lay their contributions. The publications of the Society insured their distribution to still wider circles in all parts of the world, and so led to correspondence with foreign astronomers, who gladly exchanged with our Fellows notes on points of common interest in the activities of the time.

This is naturally true of the whole history of our Society, but probably the increased facilities of communication by post and railway made themselves especially felt as the decade 1860–70 was approached. For we see signs of a distinct change in the management of our publications at that time, in the direction of increasing the importance of our shorter communications in the *Monthly Notices*, relatively to the larger papers in the *Memoirs*.

The custom of printing the octavo *Monthly Notices* and the quarto *Memoirs* has continued throughout the history. The Council considered the possibility of a departure from this custom, and decided in 1858 against making any change.

In the following year, 1859, the Council had given further consideration to the matter, and announced their change of decision in their Report on 1859 February, as follows :—

"The *Monthly Notices* continue to offer an easily accessible channel of publication to observers and computers of all classes connected with the Society, while the importance of directing the attention of Fellows from time to time to the labours of Astronomers in other countries has not been lost sight of. . . .   A plan has been carried into effect by which the *Monthly Notices* will become an integral part of the volumes of the *Memoirs*. All are aware that, for some years past, an octavo volume of *Monthly Notices* has always been given with each volume of *Memoirs* sold. It has been found that, by re-imposing the type of the *Monthly Notices* into a quarto form, with double columns, it is practicable to form an edition of the *Notices* which may be stitched up with the *Memoirs* so as actually to form part of the volume. The expense of printing the annual report of each year twice will thus be avoided. It has some- times been suggested that it was unnecessary to make the annual report a part of the volume of *Memoirs*, but those who have been students of old history have always protested against the omission. They have represented that it is a very serious defect of the older *Transactions* that they supply no materials for the histories of their several societies ; from which it not unfrequently arises that the papers themselves are unaccompanied by information necessary to their being properly understood as historical monuments. Both ends are now made to meet ; the annual report, and much current information besides, form a part of the very volume which contains the larger *Memoirs* ; and the annual report is not printed twice.''

These arrangements were carried into effect in 1860, and the publication of the *Monthly Notices* was continued in octavo form, and also in the quarto form in double columns. Thus volume **19** of the *Monthly Notices* appeared as an appendix to volume **28** of the *Memoirs* in 1860 ; and so on until 1867, when it was decided to discontinue the quarto form of the *Monthly Notices*. Thus volume **36** of the *Memoirs* is the last to contain the reimposed *Monthly Notices*, and volume **27** of the latter is the last that was reimposed. We may conclude these references to the adminis- trative side of our publications by stating that the volumes of the *Monthly Notices* for the last two years in the decade con- tained 325 and 231 pages respectively, large compared with the earliest volumes, however small in comparison with our recent volumes, some of which run to 700 or 800 pages.

Robert Grant resigned his duties as editor of the *Monthly Notices* in 1859 November, on his appointment to the Professorship of Astronomy at Glasgow. He was the author of the well-known *History of Physical Astronomy*, published in 1852, a book of per- manent value ; and to his literary tastes and his discernment in historical matters we owe a considerable debt. He introduced into our publications brief notices of valuable astronomical papers that had appeared in foreign serial publications. He was succeeded by Arthur Cayley in the editorship of the Society's publications.

Under Cayley's editorship, which continued until 1881, some convenient alterations of form were introduced in the Annual Reports. Until 1863 (in which year Cayley was elected first Sadleirian Professor of Pure Mathematics at Cambridge) there were no headings at all in those reports. Then we have headings to indicate the observatories from which reports of work had been sent, and a general heading " Progress of Astronomy "; and by 1865 we have headed subdivisions throughout the whole report. The somewhat ambitious " Progress of Astronomy " was modified in 1869 to " Notes on some points of interest connected with the progress of Astronomy during the past year." And so it has remained with but slight modification down to the present time.

An innovation was given trial during the presidency of Warren de la Rue, in the form of insertion in the *Monthly Notices* of brief reports of the discussions which took place at the meetings after the reading of the various communications. The reports were, however, not systematic, and though it was clear that such reports were regarded as desirable and likely to be of some value, they were not continued.

Possibly the improvement in the reports of discussions, which were published in the *Astronomical Register,* may have been due to this indication of an obvious desideratum. This astronomical periodical was started in 1863 January by Sandford Gorton, who was a Fellow of our Society elected in 1860. It had occurred to him that it would be very desirable " to collect together those stray fragments of information, which, though not of sufficient importance possibly to occupy the pages of the *Monthly Notices,* may nevertheless in the shape of passing conversations or occasional notes, be useful for future reference." He wished " to introduce a sort of astronomical *Notes and Queries,* a medium of communication for amateurs and others." It aimed further at giving an account of the discussions which took place at the meetings of the Royal Astronomical Society, both for the sake of those who were unable to be present and also in order that some permanent record of them should be preserved.

It must be admitted that the reports of the discussions were of a very slight description at first. One would judge that very often they recorded only some pithy remarks separated from their serious context, and giving but a poor idea of the discussions in which they were let fall. One can hardly imagine that the real gist of the Astronomer Royal's remark about Bessel's probable error in his measures of the parallax of 61 Cygni are justly recorded when he is reported as saying that " these probabilities are not worth a pin ! " However, as time went on, the discussions were better reported, and are often interesting as the only record available.

The editor seems frequently to have had considerable difficulty in satisfying the desires of all his various readers with regard to the correspondence admitted to his columns. The great controversy about the lunar rotation, in which Henry Perigal was one of the protagonists, went on for months with persistent reiteration of misunderstandings. Even Augustus De Morgan was drawn into it, and he was reproached by one of the correspondents for that in answer to a demand for a proof of rotation, he intimated that a proof demands a capacity for the reception of proof.

The pages of the *Register* bear testimony to the interest taken in a wide circle of amateur observers in such debated matters as the question of variation in the lunar crater Linné, when Schmidt, of Athens, called attention to it; and the telescopic appearances presented by the sun's surface when Nasmyth announced his view of interlacing willow leaves, and a fierce battle arose as between granules, soapsuds, and even cauliflower heads.

There can be no doubt that the *Register* was very welcome to many observers who desired to have their work usefully directed. It became a means of publication of the earliest reports of the Observing Astronomical Society, which was started in 1869 under the keen Secretaryship of Mr. W. F. Denning, who was then in his twenty-first year. In the following year the Society contained forty-six members, and it may be regarded as an early forerunner in an aim which later found a really fine fulfilment in the foundation of the British Astronomical Association in 1890.

The Rev. J. C. Jackson became editor of the *Register* in 1872, when Gorton's health failed. The publication was continued to 1886 December, and volume 24 was completed with the shortest editorial note, " Finis, Valete."

In the attempt to prepare a chapter of history like the present, a feature that strikes the compiler is the great value of the records stored up each year in the Annual Reports of the Council. The Addresses of the Presidents in the awards of the medal, the Council Notes on Points of Interest, and the Obituary Notices, all combine to present a view of the activities of any epoch in a way that is rendered all the more valuable, inasmuch as the part played by any particular research or by any individual is presented at different times, and thus our final view of it is modified by the very fact that we have, firstly, the Note calling contemporaneous attention to a point of fresh interest and importance ; secondly, possibly an Address setting forth a later view of a specially selected piece of work in its relation to astronomical progress ; and thirdly, the Obituary Notices, affording generally a still more distant judgment of the contributions of individual workers. Indeed, the abundance of records is in some ways even an embarrassment ; for the compiler is in danger of becoming interested in points on which he feels he should enlighten his own ignorance. Effort has been made as far as possible to let the records speak for themselves, and tell the tale of the decade.

After these introductory remarks we take up the true theme of this chapter, the history of the Society in this last decade of the first half-century of its existence.

The decade began under the presidency of the Rev. Robert Main, F.R.S., who had been elected in 1860 February for the second year of his term of office. He was then completing the twenty-fifth year of his activities as Chief Assistant at the Royal Observatory, Greenwich, an office to which he had been appointed by Airy in 1835, when he succeeded Pond as Astronomer Royal. Main had been a very faithful officer of the Society, and after five years as one of the Honorary Secretaries, 1841-46, the Council made a warm acknowledgment of his services. He had contributed many important papers to the *Memoirs*, and the value of those contributions to the promotion of Astronomy had been recognised by the Society in the award of the Gold Medal to him in 1858. Main was evidently greatly respected by his contemporaries as one who, quite apart from his devotion to his own immediate work, spared himself no trouble in arriving at sound judgments of the value of astronomical investigations within his cognisance. He delivered three addresses in setting forth the grounds of the award of the Gold Medal in successive years; firstly, to Carrington, for his Redhill Catalogue of stars within 10° of the Northern Pole of the heavens ; secondly, to Hansen, for his Lunar Tables ; and thirdly, to Goldschmidt, for his discoveries of thirteen small planets.

Main's address in 1860 on Hansen's Lunar Tables was a long one ; and it has a special value. It gives both a summary of the early work on lunar observations and theory, and also a weighty indication of the contemporaneous view of the great value of Hansen's work.

Main's third address, in 1861, on Goldschmidt's discoveries of minor planets, reminds us of the value of work done by an amateur in another country. Goldschmidt was an artist living in Paris, and had passed the age of forty-five before the accident of hearing a lecture at the Sorbonne by Le Verrier, in which he called attention to an eclipse of the moon that was to occur on the same evening, aroused in him an enthusiasm for astronomical study.

Two years later we learn, with the proceeds of the sale of one of two copies he had made at Florence of a portrait of Galileo, he procured a small telescope of about two inches aperture. With the help of the Berlin Star Charts he discovered his first small planet in 1852, and Arago named it Lutetia. With slightly increased telescopic power, in the next nine years he had discovered twelve or thirteen more. He was elected an Associate in 1866, and died only a few months later at Fontainebleau.

The rapid increase in the number of discoveries of minor planets led to an agreement in 1863 between the Observatories of Greenwich and Paris for a distribution of the labour of the meridional observations of these small bodies. The agreement took a peculiar form, which was determined by the obligation of the Royal Observatory to maintain meridional observations of the moon, a matter which had always been of high importance in the responsibilities of Greenwich. Airy and Le Verrier arranged to divide the additional labour of observations of the planets, by the agreement that the Observatory of Paris should undertake them from full moon to new moon, the Observatory of Greenwich remaining charged with those from new moon to full moon. Then was inaugurated what has been claimed as the first specific plan of co-operation among astronomers. In the following year the Director of the National Observatory, Washington, also promised to co-operate in the observations.

Early in 1861 the Treasurer, Whitbread, had called the attention of the Council to the fact that the yearly expenditure had exceeded the annual income by over £200. A Committee was appointed at once to examine into the general subject of the income and expenditure of the Society. They proceeded in the most business-like fashion to their task, and drew up a valuable report. It appeared from it that the average expenditure of the preceding four years had been about £60 in excess of the average income, which was increasing only at the rate of about £18 a year. The adverse balances were ascribed as principally due to the growth of the bills for printing. With respect to the treatment of compounders' fees, the Committee pointed out that :—

> By a minute of Council, March 1820, it was resolved that all compositions should be funded, and the interest of the fund alone treated as income.
>
> By the minute of June 1828, it was recommended that on the decease of any compounder his composition should, if needful, be made available for general purposes, but that the permanent fund should never be reduced below an amount equal to the product of £21 by the number of surviving compounders.

By the minute of June 1857, the Council resorted to an intermediate measure, and left it to the Treasurer to advise every half-
year. The Committee further stated that the funded property of
the Society, excluding the Lee and Turnor funds, amounted to
£4900, and that the number of compounders was 160, whose
compositions were represented by £3500 ; and that the position of
the Society was therefore more than solvent in this respect, there
being an excess of £1400, accumulated partly by bequests and partly
by saving. They concluded their report by certain recommendations, dictated, they said, rather by motives of policy than by
necessity :—

(a) That as nearly as circumstances will allow, all compositions
should be funded.

(b) That considering that the *Monthly Notices* have now attained
a bulk amply sufficient for their intended purpose, the editor be
desired not to exceed 24 octavo sheets.

(c) That in the case of papers for the *Memoirs*, the actual ballot
be deferred to the meeting of the Council in June of each year, so
as to allow of the formation of a scheme for the whole volume for
the year.

These recommendations, with a couple of others of minor
importance, were unanimously adopted, as appropriate for immediate action.

Samuel Charles Whitbread [1796-1879] had joined the Society
in 1849, and succeeded George Bishop as Treasurer in 1857; he
reigned over our finances for twenty-one years. Whitbread was
M.P. for Middlesex for ten years, and in spite of his interests in
politics and hunting, he found time in which he devoted himself
to the study of astronomy and meteorology, building an Observatory at his residence at Cardington, near Bedford, and becoming
with John Lee and James Glaisher one of the three founders of the
Meteorological Society in 1850.

In 1861 a Committee, consisting of the Astronomer Royal,
Manners, Vignoles, Adams, Whitbread, Jacob, De la Rue, and
Carrington, was appointed to take into consideration the advisa-
bility of establishing for a limited number of years a Hill Observatory in India. The matter had been mooted and much discussed
in 1858-9, but it had been laid aside in consequence of the unrest
which followed the Indian Mutiny. The subject was revised by
Carrington and Jacob, who had recently resigned the Directorship
of the Madras Observatory by reason of ill-health. Jacob submitted his views to the Committee in the following terms :—

It has constantly been remarked that it would be indeed
difficult among numerous observing stations and fine instruments
which now exist, to point to a single one where a telescope of decent,

respectable quality and power was placed in a position where its capabilities could be brought out and the utmost obtained from it. . . . It has become more than ever apparent that size is not so much the quality to be sought as exquisite definition, excellence in which requires admirable workmanship first and an admirably pure, tranquil and continuously transparent sky afterwards.

The programme of observations included the following objects as desirable : Physical Observations of planets and satellites, especially Mars in 1862, parallax of Mars, observations of nebulæ, variable stars, zodiacal light, and double stars.

The final proposal submitted by the Committee was " to place an equatorially mounted refractor of not less than 9 inches aperture and of high optical excellence in the charge of Captain Jacob, at a station to be selected from the many accessible points in the neighbourhood of Poona, some eighty miles from Bombay. There an elevation of some 4000 feet could be obtained " ; and the proximity of arsenal and artificers would naturally be convenient.

The Council approved of the recommendation, and decided in 1861 June to make application for the aid of Her Majesty's Government towards the establishment for a limited period, under the superintendence of Captain Jacob, of an Observatory in the neighbourhood of Bombay, at a considerable altitude above the sea. They received the following letter, which deserves to be recorded once more as an instance of support promptly given by the Government to an astronomical enterprise :—

TREASURY CHAMBERS,
8 *August* 1861.

In reply to your application addressed to Lord Palmerston on the 24th June last, for a grant to the Royal Astronomical Society of £1000, in aid of the proposed temporary maintenance of an observatory near Poona, I am commanded by the Lords Commissioners of Her Majesty's Treasury to acquaint you that the sum of £1000 having been voted in Parliament for the object described in your letter, My Lords will be prepared to issue the amount in such manner as you may desire, on the understanding that the Society will see to the proper application of the fund thus placed at its disposal.—I have the honour to be, Gentlemen, your obedient servant,   (Signed)   GEO. A. HAMILTON.

To the President and Treasurer of the
Royal Astronomical Society, Somerset House.

A letter was sent to his Lordship expressing the thanks of the Council for the promptness with which their application had been met. The sum granted was immediately paid to the account of the Society, and Captain Jacob, having purchased at his own expense a telescope of aperture 9 inches from Messrs. Cooke, of York, sailed on 1862 April 26, with his wife and family, and instruments, and landed at Bombay on August 8. A letter to Piazzi Smyth ends with a postscript dated August 11 : " Leaving for Poonah to-morrow. All Well ! " This was followed by the news of his death on the 16th. And so in this tragical disaster ended an enterprise which had started with so great promise of success.

The Society has always been proud that Her Majesty Queen Victoria had, on accession to the throne, graciously complied with the request that Her Majesty should become the Patroness of the Society. It is an indication of the widespread sympathy and anxiety on account of the illness of the Prince Consort, that the Council Meeting which had been called for 1861 December 13, was adjourned after the only business explicitly prescribed by the bye-laws had been transacted. The Prince died on the following day, and the Council met again on December 18 and drew up the address of condolence, which was submitted to the Society on 1862 January 10 for presentation to the Queen.

The second year, 1861, of the decade began with an episode which roused strong feeling at the time. It related to the election of a new President to succeed Robert Main. In 1858 the Council had reported that it had been invited by the united request of five of the Fellows to discuss an alteration in the mode of electing the Officers and Council. The proposed change seemed to arise out of the opinion that the then existing method, which consisted in bringing to the vote a list prepared by the retiring Council, with individual liberty of substitution of any one name for any other, gave no opportunity of previous concert in the election of officers, except among those Fellows who happen to be thrown together by circumstances. The matter was referred to the next Council, and the result was that a Special General Meeting was held on 1858 June 11, and alterations in the Bye-laws were enacted, by which the practice of submitting a list of Officers and Council for election, hitherto followed by the Council for convenience, was enjoined by a bye-law. But, in addition to this list, any names forwarded by any two or more Fellows before the ordinary meeting of the Council in December, were to be submitted to the General Meeting in February : the common right of striking out any of those names and substituting others remaining unaltered. The lists were to be circulated as soon after the meeting of Council in December as could be conveniently done. The effect of the change was frankly described (by the Council which retired in 1859 February) as being that a much longer time

A. DE MORGAN
(1806–1871)
(*By kind permission of Messrs. Speaight Ltd.*)

To face p. 140.]

is given for deliberation and concerted action ; and any two Fellows could make known their joint opinions as to the persons most fit to be the Officers and Council of the Society without trouble or expense to themselves.

For the election of Officers and Council in 1861, the Council received from six Fellows a requisition nominating John Lee, of Hartwell House, for the office of President ; the Council's choice for President was Airy. Accordingly the list was printed and circulated, in accordance with the new Bye-laws, in a form showing two nominations for the Presidency.

The ballot resulted in the election of Lee as President and Airy as Vice-President, and the rest of the list as proposed by the Council.

It is not easy now to understand the feeling in the Society that could have allowed the recommendation of the Council to be overridden. Airy had, it is true, already served three times as President. Lee was completing his 78th year at the time of his election.

De Morgan, who had the welfare of the Society very much at heart, as is well shown by his two terms of eight years as Secretary and long service on the Council, could look at the humorous side of the matter in writing privately to Airy : " It is wondered that the *Airy* party, who must have had the wind, should have allowed the Society to fall to Leeward." But he took a very different line in letting the Society know what he thought of their action. He wrote to the Council, saying that he would not accept the office of Vice-President, and requesting that his letter should be laid before the Society by being entered on the Minutes. It seems a long and laboured indictment, and few men would either press the analysis of the motives of their action to their bitter conclusion as De Morgan has done, or care to have it published. After much consideration as to whether it should be recorded here, it has seemed right not to print it. Mrs. De Morgan published it in full in her *Memoir of Augustus De Morgan.* London, 1882, pp. 272–278.

In looking back now at the episode in the absence of any contemporaneous criticism, it almost seems as if De Morgan, in his warm advocacy of the traditional custom by which the Society had hitherto always followed the lead of the outgoing Council in the nomination of the incoming Council, may have misjudged the situation. The action of the Society is possibly to be attributed not so much to a wish to upset a tradition, but rather to a chivalrous desire to recognise Lee's frequent readiness to act as Chairman during the Presidency of George Bishop, who almost simultaneously with his election as President in 1857 had been overtaken by illness, and was never once able to preside at a meeting of the Society. Airy, Main, and Baden Powell each acted as Chairman three times ; and Lee, six times.

De Morgan's friends naturally besought him to remain upon the Council. He was, however, firm in his resolve, and he never served again upon the Council. When, however, towards the end of the year he was asked by the Council to lend them his friendly services, as in past years, in the preparation of the Annual Report, he undertook the work and carried it through. His combination of rigid principles with good nature and sense of humour is well known to those who are acquainted with his " Budget of Para- doxes " ; and we could not have anticipated a refusal to help in a good cause from a man who, when called upon to defend the De in his name, could say that as he had seen in catalogues his own name between those of De Moivre and De Mosthenes, he was constantly tempted to make the same mistake in the Greek name.

John Lee (1783–1866), who was thus elected President in 1861, was the eldest son of John Fiott, and of Harriett, daughter of William Lee, of Totteridge Park, Herts. He had graduated at Cambridge as fifth wrangler in 1806, and was elected to a Fellowship at St. John's College. Having obtained a Travelling Scholarship, he visited Greece, Egypt, and the Holy Land, and amassed a valuable collection of antiquities. His antiquarian studies seem to have been the main interest of his life. In 1815 he had assumed the name of Lee by royal licence, in compliance with the will of his maternal uncle. On the death of Sir George Lee, Bart., in 1827, the whole of the family property devolved upon Dr. Lee, and he then became Lord of the Manors of Hartwell, Stone, and Bishopstone, and patron of two livings, to which reference has been made in a previous chapter. At Hartwell House he had an observatory built, where for many years astronomical and meteoro- logical observations were carried on.

Admiral W. H. Smyth (1788–1865) was an enthusiastic instigator of Dr. Lee's astronomical efforts. He was the son of an American Loyalist, entered the navy in 1805, and was actively engaged until 1815 in the Indian Seas and on the coasts of Spain and Italy, where he had his full share of adventure and danger. From 1817 to 1824 he was engaged in the great survey of the Mediterranean, which has been described as " the greatest scientific survey ever planned and completed by one individual." After the end of his naval career he settled at Bedford in 1828, and there his duties as a magistrate brought him in frequent association with Dr. Lee. A great friendship arose between them, stimulated by common interests ; and their mutual influences made Smyth an antiquarian, and Lee an amateur astronomer. There is a human touch in the description of the meridian marks which Smyth had made for Lee's transit instrument. The adjustable metallic marks were let into blocks of marble, the northern block being a representation of the Temple of Janus, as given on a large brass medal of Nero, whilst the southern block was a miniature of the façade of the Temple of Concord at Girgenti, with its central columns omitted for the insertion of the meridian plate.

Smyth had moved from Bedford in 1842 to St. John's Lodge, within a short walk from Hartwell House. Lee purchased his instruments, and the two friends pursued their astronomical studies together. Dr. Lee employed James Epps (1771–1839) and Norman Pogson (1829–91) as assistants, Pogson coming in 1859 January from the Radcliffe Observatory at Oxford, and remaining at Hartwell till he was appointed Government Astronomer at Madras in 1860 October. Pogson's work at Hartwell, relating mainly to the study of variable stars after the method of Argelander, was directed to the formation of an atlas of variable stars, and he had completed nineteen charts, 80' square, on a scale of 3 inches to a degree, with stars noted down to the twelfth magnitude, and with accurate magnitudes of certain comparison stars. Pogson carried on the work at Madras, and it appears that out of the nineteen charts and catalogues reported by Pogson as completed in 1860, six had been engraved and printed for Dr. Lee ; for copies of them were found by Dr. Copeland in the library of the Edinburgh Observatory. But Pogson's systematic work had to wait till 1908 before it was published, after being prepared for press by the volunteer labour and helpful generosity of Mr. C. L. Brook, with an illuminating introduction by Professor H. H. Turner, who has done so much to save from oblivion vast materials of early observations relative to variable stars. To Pogson we owe the discovery of many variable stars, and to him we owe the suggestion of the definition of the magnitude relation—2·512 times the logarithm of the brightness—a relation that lies at the foundation of all modern photometric work.

In these last paragraphs we have an instance of the difficulty which the history of the Society imposes on a writer charged with a single decade. He is diverted from a record of a single President into an inadequate reference to work upon variable stars. But still it is possible thereby to indicate the kind of service the Society is able to render in a single branch when individual members are willing by their loyalty and their labours to advance the cause of Astronomy. We see the influence of Dr. Lee, a generous patron of Astronomy,—stimulated by that outspoken free lance, or perhaps we should better employ Dr. Lee's own term, " that indefatigable officer," Admiral Smyth, who was always having his tilt at the " magnates and dons "—maintaining a private observatory, chosing a promising line of research then in its infancy, leaving it in the hands of a capable and enthusiastic observer, initiating work that only comes to fulness years later under the fostering help of the Society, aided by its loyal members.

Lee delivered Presidential Addresses in presenting the medal to De la Rue in 1862 for his astronomical researches, and especially for his application of photography, and to Argelander in 1863 for his survey of the northern heavens.

Of De la Rue's work the Council in their report could " not help remarking that this public recognition of the success of chemical delineation of celestial objects may be an important date in the history of Astronomy. No discovery of our day affords a more hopeful field of anticipation than that of photography, which seems destined to take that part in the astronomy of visual phenomena which graduated instruments have taken in the Astronomy of motions and positions."

The antithesis suggested in this quaintly worded comment seems to have been dictated by the recognition that De la Rue, stirred to emulation by Bond's daguerreotypes of the moon taken at Harvard College Observatory in 1850, applied the newly invented collodion wet plates to the photography not only of the moon, but also of the sun, the prominences and lower corona, and of Jupiter, Saturn, and some double stars, but had not yet fully utilised his records for measurement.

Lee, in presenting the medal to him, justly said that " his claim does not rest on the absolute priority of his application of a well-known art in a new direction. It is rather based on the fact that by methods and adaptations peculiarly his own, he has been the first to obtain automatic pictures of the sun and moon, sufficiently delicate in their detail to advance our knowledge regarding the physical character of those bodies, and admitting of measurements astronomically precise."

We shall have more to say about De la Rue's organisation of measurement, and his other astronomical work, when we come to deal with his years of Presidency.

Of the award to Argelander, the Council's Report (1863) makes a bare announcement. Dr. Lee, in his address on the occasion, began by giving a short retrospective view of astronomical achievements since the foundation of the Society, and refers to the balloon ascents of Glaisher and Coxwell. In this connection a note in the Council's Report for 1862 may be quoted :—

One of the main objects of these ascents was to extend and improve our knowledge of the relation which exists between increase of elevation and the corresponding variations of temperature and moisture, these variations in their turn having an intimate bearing on the theoretic determination of atmospheric refraction.

The results of Mr. Glaisher's observations indicate that the hypothesis of a diminution of 1° Fahr. of temperature for every additional 300 feet of elevation must be abandoned even for inconsiderable heights above the earth's surface. . . . Through the first 1000 feet under a cloudy sky there is a change of 1° Fahr. for every 213 feet, but if the sky be clear there is a diminution of 1° Fahr. for every 139 feet. At extreme heights, such as between 25,000 and 30,000 feet, it requires a full 1000 feet of additional altitude to cause a diminution of temperature of 1° Fahr.

Herein is found the first discovery of indications of what in modern days is recognised as the stratosphere.

It is difficult to see how Lee, in speaking in 1863 of Argelander's achievement, " his Survey of the Northern Heavens," could have missed the opportunity of a fine address. It would almost seem as if he could never have had access to any of the three volumes, 3, 4, and 5, of the Bonn Observations, which constitute volumes I, II, and III of the Bonn Catalogue, though they bear the date 1859, 1861, and 1862 respectively. He appears rather to have contented himself with the slight information contained in two notes published in the *Monthly Notices* : the first being a report of progress of the charts communicated to Carrington by Krüger at the request of Argelander in December 1861 (*M.N.*, **22,** 57); and the second being a note of twenty-six lines in the Council's Report in 1862 February (*M.N.*, **22,** 125).

Lee would seem even to have read Carrington's comment on Krüger's note unmoved. It deserves to be transcribed here :—

The Fellows of the Society, and all to whom this account comes, cannot fail to admire the stately progress of this enormous work, which, simple in its conception, and free from insurmountable difficulties in execution, would appal many an astronomer by its vast extent.

The three volumes of the catalogue contained the magnitude, R.A. and Dec. reduced to epoch 1855·0—approximately the mean date of the zone observations completed between the years 1852 and 1859—of 324,198 stars. The catalogue was completed by the issue of an atlas of forty charts containing a representation of every star in the catalogue; thirty-seven of them had been published at the date of Lee's address, and the forty were complete by the summer of 1863.

The value of the work as a preliminary survey of the Northern Heavens—not of high precision but of enormous utility for the identification of stars to the ninth magnitude—has received full recognition from grateful astronomers, who to this day plot fields of stars from Argelander's charts and use Argelander's sign manual, such as B.D.+31° 2327, in their constant references. His assiduous assistants Schönfeld [1828–91] and Krüger [1832–96] are not forgotten.*

That this work was at once recognised as worth the trouble spent upon it was proved by the continued efforts to secure a similar survey of the Southern Heavens. In 1862 our Council, on the initiative of Carrington, who was doubtless moved by the publication of Argelander's *Durchmusterung* volumes, appointed a Committee (Airy, Carrington, and Hind) to report on the best means to be pursued to carry out a survey of the stars of the Southern Hemisphere. The Committee suggested the Cape, Sydney, Williamstown (Melbourne), and Hobarton, as possibly providing men willing to undertake the work. It would require six or eight years to record 300,000 stars twice and to revise ; a special staff would be needed, but the cost of the instrumental requirements might be taken as negligible. In 1863, Strange's name was substituted for Airy's, and a few months later the names of Pritchard, Hodgson, and Manners were added. Meanwhile enquiries had been made of Maclear at the Cape, Pogson at Madras, Smalley at Sydney, Ellery at Melbourne, and Todd at Adelaide, resulting in Maclear's accepting the polar segment from the south pole to declination − 50°, and Ellery's undertaking the zones between declinations − 20° and − 40°. In 1864, Adams was appointed to the Committee in place of Hind, and in 1866, Stone, who also had joined the Committee, was instructed to write and express satisfaction at Ellery's progress. But the undertaking languished. And though, as will be later recounted, at Argelander's instigation the huge undertaking of the Astronomische Gesellschaft was initiated in 1867, the Southern Heavens had to wait for their complete survey until Gill and Kapteyn carried out the photographic survey in the years 1893–1901. Meanwhile, Schönfeld, who had left Bonn in 1859 for Mannheim, but took much of the Bonn work with him to complete it, had returned to Bonn as Director of the Observatory on the death, in 1875, of his former chief and life-long friend, Argelander. He at once embarked upon his extension of the survey as far as declination −23°, and by 1886,

* Richard Proctor (1837–88) made an interesting chart by laboriously plotting every star of Argelander's list on a single circular map of the Northern Hemisphere on an equal-surface projection. Many of his earlier books, which did much to spread an interest in Astronomy, were published in this decade.

after ten years' work, his catalogue of 133,659 stars was completed and published in volume eight of the Bonn Observations.

In 1863, Airy [1801–92] was elected to the Presidency for the fourth time, in spite of his expressed desire to be excused from office. He served for a single year, and he was relieved of the labour of preparing an address at the Annual Meeting of 1864; for no medal was awarded. The Vice-President (Main), however, gave a short address, and stated that the decision of the Council not to adjudge the medal was occasioned not by any want of meritorious work which deserved such a reward, but rather on account of the very abundance of them. Main called attention to various points mentioned in the Council's report, and added that " in the account of the progress of English Astronomy it would be unpardonable not to give a prominent place to the publication of Mr. Carrington's book on the Solar Spots. It would be premature to eulogise this admirable production, the merits of which would, no doubt, at some future time be brought prominently before the Society; but he could not avoid saying that he recognised in it the same careful elaboration and finish, the same attention to the minutest necessary details of observation and calculation, coupled with the broadest and most sagacious theoretical views which had rendered the Red Hill Catalogue of stars a classic among similar productions."

Main had given the address when the medal was awarded to Carrington for the Red Hill Catalogue in 1859, and the anticipation which the words just quoted suggest that a similar award might soon be made for the solar work rouses some wonderment as to whether the Council in 1863 November, having thought of awards to several Astronomers, had refrained out of consideration for Airy. It is on record that Airy had asked the Council to relieve him of the obligation to prepare an address; and if we are left with a feeling of surprise that Carrington's solar work was never crowned, it may be stated that it is also on record that Carrington, having learnt that the Council in 1865 were preparing to include his solar work in the list of possible awards, requested them to omit his name from the list.

That Airy had reason to hesitate about undertaking any work beyond his onerous official duties, we in these days have learnt from the records of his activity published after his death. Main had given some indication of that activity when he was speaking of the help he had received from Airy in the preparation of the address on Hansen's Lunar Tables in 1860. He described his amazement at realising for the first time the amount of anxiety and labour which had fallen on Airy, though he (Main) had been for nearly five-and-twenty years so near his person, and generally

cognisant of all his efforts for the advancement of lunar theory
(*M.N.*, **20**, 166).

> I was not aware myself of the amount of anxiety which the
> bad representation of the lunar motions by the Tables of Burckhardt
> had caused, more than twenty years ago, in the minds of some of the
> greatest of our astronomers, till, when I was preparing for my
> present task, the Astronomer Royal, with his usual kindness, put
> into my hands his correspondence on lunar theory, continued from
> the time when he was first called upon to occupy his present position,
> namely, the year 1835, to the present time. Even if I had not known
> how many great works he has brought to completion in the same
> period—how he has revolutionised the theory and the practice of
> the construction of astronomical instruments, as well as of the
> making and reducing of observations—how he has borne the chief
> labour in almost every Government commission for scientific
> purposes, of which we need only mention the Standards Commission
> —how he has occasionally been engaged in optical researches, or
> in the writing of profound memoirs in some branch of abstract or
> mixed science, some of which adorn our own Transactions—and all
> apart from and in addition to the direct duties of his office—I should
> have thought the amount of care and labour which this subject of
> the lunar theory brought upon him, even before he took upon himself
> the responsibility of the reduction of the ancient Greenwich Ob-
> servations, a very serious addition to the heavy duties which are
> necessarily imposed upon him by the ordinary administration of
> the observatory.

The reductions of the Greenwich Lunar Observations 1750–
1830 were published in 1848, and the supplementary work bringing
the reductions complete to 1851 was published in the summer of
1859. Main's words serve to indicate the respect in which Airy's
activity was held at the beginning of the decade with which we
are here concerned. His work enabled astronomers to rise above
the inadequacy of the Lunar Tables of Burckhardt [1773–1825],
and prepared the way for the great advance made by Hansen
[1795–1874]. The fruits of his labour were being gathered at the
beginning of this decade. It must have been an immense satisfaction to Airy when Hind, the Superintendent of the Nautical
Almanac Office, was able to give him the results of the computations of the moon's places from Hansen's Tables complete for the
years 1847 to 1858. Adams's discovery of errors in Burckhardt's
treatment of the moon's parallax enabled Airy to give a rigorously
fair comparison of Burckhardt's and Hansen's Tables with the
Greenwich Observations, conclusively in favour of the accuracy
of Hansen's. "Probably in no recorded instance has practical
science ever advanced so far in accuracy by a single stride."

And so it came about that Hansen's Tables held sway for a generation.   Of Airy's own later work this is not the place to speak.

Meanwhile new activities had grown up in Airy, and he strove through the Society to promote co-operation among astronomers on another problem of prime importance, the solar parallax. The inadequacy of Encke's value, 8″.57, which was adopted in 1824 and was still generally in use, had been indicated by Hansen's, Le Verrier's, and Foucault's researches ; and Airy said that new observations could alone help us satisfactorily to correct our obviously inadequate knowledge of the sun's distance.   As early as 1857 he passed in review the methods " available for correcting the sun's distance in the next twenty-five years " ; and while laying special stress on the transits of Venus in 1874 and 1882, he pointed out the desirability of taking advantage of the favourable oppositions of Mars in 1860, 1862, and 1877.

But activities of this kind were the essence of Airy's life, and this is not the place to follow them up in detail.   It is enough to say that the influence of his personal and official labours contributed enormously to the regard in which the Society was held both at home and abroad.

Simon Newcomb [1835–1909], who in his *Reminiscences of an Astronomer*, published in 1903, has recorded his experiences in visits to England and elsewhere in Europe—his first visit to Greenwich was in 1870—gives what one may regard as a contemporaneous view of Airy.   " We may look back on Airy as the most commanding figure in the astronomy of our time.   He owes this position not only to his early work in mathematical astronomy, but also to his ability as an organiser. . . .   He introduced production on a large scale into astronomy."

The problem of determining the movement of the Solar System in space was brought into prominence again in this decade by the increase in our knowledge of the proper motions of the stars. Airy introduced his new method of dealing with such proper motions, in a paper published in the *Memoirs*, **28**, 143, dealing with 113 stars.   Sir William Herschel's method, based upon an incorrect assumption that the brighter stars are in general nearer to us than faint stars, was unsuitable for the discussion of proper motions of a great number of stars, though it was well adapted to the limited facts of observation known to him.

Airy sought for a method which should not, like Bessel's and Argelander's methods, depend upon a knowledge of an approximate position of the apex of the solar motion.   He accordingly decided that instead of using the apparent angular motions of the sun and stars as exhibited on the surface of a globe, it would be preferable to treat their linear movements by reference to rectangular co-ordinates with the sun as origin and the pole on the axis of Z. He illustrated the application of his method by treating the proper motions of 113 stars, taken from Main's recent determinations, and deduced an apex of the sun's motion on each of two suppositions; first, that the irregularities of proper motions are entirely due to chance error of observations; and second, that such irregularities are entirely due to the peculiar motions of the stars. A couple of years later, Dunkin had applied Airy's method to the discussion of the proper motions of 1167 stars from Main's values; and the results gave positions of the apex, both for Airy's 113 stars and for Dunkin's 1167 stars as follows :—

		Airy, 113 Stars.		Dunkin, 1167 Stars.	
1st Supposition	R.A.	256°	D. +39°·5	261°	+39°·5
2nd Supposition		261°	+24°·7	263°	+25°·0
To compare with					
Herschel		260°	+29°·3		
Argelander		256°	+38°·6		

Airy recognised, of course, that his method, as then applicable, had to proceed on the incorrect assumption that the stars were all at the same distance from the sun; and he expected to find that light would be thrown upon the matter by the way in which the sum of the squares of the residual errors was affected. But no definite conclusion was forthcoming, "I therefore asked Mr. Stone to examine the matter, as I may say, *maliciously*, to discover if there were not some error : he has gone into it and can find nothing wrong. Supposing Bradley made errors of right ascension, that might account for a good deal; but the matter is left in a most delightful state of uncertainty, and I shall be very glad if anyone can help us out of it."

Much water has flowed under the bridges, as the saying is, since then. There has come an immense development of statistical methods in the last twenty years. But the episode is peculiarly characteristic of Airy's philosophical mind.

To the year 1863 belongs also the beginning of another centre of co-operation, the birth of the Astronomische Gesellschaft, which was formed at Heidelberg on August 28, with a Council containing the names of Argelander, O. Struve, Bruhns, Schönfeld, Zöllner, Zech, and Foerster. With the death of the last-named, in 1921 January, the last personal link with the original founders is gone.

De la Rue, in addressing the Society after the recess in the first year of his Presidency, 1864, gave a very interesting account of a visit paid by him to Pulkowa, in compliance with an invitation which he had received as President of our Society, to take part in the festival given on the twenty-fifth anniversary of the inauguration of the Central Observatory of Russia. He had, he said, been the more ready to accept the invitation because he had been informed that many of the members of the Council of the recently founded Astronomische Gesellschaft were to be present, and he anticipated that it would be agreeable to them to meet some officer of the elder English Society with a view to the organisation of a friendly alliance, if not of a more intimate connection, between the two Societies. He explained that the objects of the new Society were in great measure different from those of our own Society.

> In the first place, it is not intended that it shall meet periodically to receive and read communications on astronomical subjects, and it does not contemplate the publication of *Memoirs* or *Proceedings* in the ordinary meaning of those terms. It has no honorary members, no corresponding members or Associates, and is, in fact, more properly a co-operative union of astronomers, who propose, either by correspondence or by the occasional assemblage of its members in various towns, to effect an interchange of ideas, and to promote and encourage, by concerted and well-directed action, such undertakings as may, from time to time, appear best calculated to aid the progress of astronomical science. Our German colleagues have long felt the necessity that some understanding should be come to among scientific men to prevent a waste of valuable activity from the circumstance of two or more astronomers working on the same subject ; also that the zeal of the numerous contributors to astronomy would be rendered much more effective if it were in some measure directed by a concerted action. It is thought that in many cases much good may be done by causing certain preliminary investigations to be made, and, in others, by effecting a revision and new reduction of the older observations (Bradley's, for example) ; hence it is contemplated to use the funds of the Society in getting this work done when necessary by paid computers, and it is their further intention to print and issue as speedily as possible the results obtained.
>
> It will be a matter for consideration in what way our Society can ally itself with the German Society, and it may not be very readily seen how this can be accomplished, but I am sure you will concur with me in wishing most cordially that that Society may prosper, and possibly many of you will like to do as Professor Adams and myself have done, join the Society and become entitled to its publications.

The earliest biennial meetings of the Astronomische Gesellschaft were held, the first at Leipzic in 1865, the second at Bonn in 1867, the third at Vienna in 1869; and the *Vierteljahrsschrift* was started in 1866. By the end of the decade the co-operative plan for the accurate determination of the places of all stars to the ninth magnitude between −2° and +80° declinations was organised by the assignment of the separate 5° zones to various observatories in Europe and America. Nearly ten years passed before the work could be regarded as started in the majority of the zones; twenty years elapsed before the first instalments of the *Catalogue* were published. The issue in 1910 of the *Berlin C. Catalogue* for the zone +70° to +75° declination—a zone originally assigned to Dorpat—completed the plan contemplated at the time of the inception in 1869.

Another piece of work, which was put on the Astronomische Gesellschaft's list of desiderata in 1866, was a new reduction of Bradley's Observations. It was undertaken by Auwers in 1866 and completed in 1876. The results of his laborious undertaking were published in three volumes, which were issued in the years 1882, 1888, and 1903.

It was in 1863 that Huggins and Miller began to publish their spectroscopic investigations of the chemistry of the stars, investigations which had been instigated by Bunsen and Kirchhoff's researches in 1859–60. Our *Monthly Notices* for 1863 March contain a reference over the now well-known initials J. N. L. to a brief note communicated by Huggins and Miller to the Royal Society in February "On the lines in the spectra of some of the fixed stars." This note was a very brief forerunner of their first Memoir in the *Philosophical Transactions*. Lockyer's reference to it is of interest, as showing at this early date the difference in trend of thought in the new subject of stellar spectroscopy, a subject in which Huggins and Lockyer came in later years to be almost as frequently in disagreement as in agreement. Lockyer's references to the note in question showed full appreciation of Huggins's method of direct comparison of terrestrial and stellar spectra; he saw, too, the possibilities opened out of gauging the temperatures of the sun and stars, and utilised Bunsen's estimate of the temperature of the oxyhydrogen flame to hazard a guess at a similar temperature for the sun. He added that Huggins and Miller had succeeded in obtaining microscopic photographs of the spectra of Sirius and Capella.

In the following number of the *Monthly Notices* (**23**, 188), Airy gave a brief note describing the experimental apparatus prepared at the Royal Observatory for the observation of stellar spectra. In this apparatus, ingenious use is made of focal lines formed when an uncollimated beam falls on a prism displaced from the position for minimum deviation. The work was initiated under the charge of James Carpenter, who later collaborated with James Nasmyth in the work on the moon (1874).

SIR WM. HUGGINS
(1824–1910)

To face p. 152.]

Here were the beginnings of the introduction of precise spectroscopy into astronomical investigations.

Both Huggins and Lockyer, who stood out as the protagonists in the new methods, communicated most of their work to the Royal Society, though the fact that Huggins was one of the Secretaries of our Society from 1867 to 1872 kept him in very close touch with our Society, and he was frequently called upon to speak of the nature and real significance of the harvest of data that were being gathered in the new branch of astronomy to which he devoted his pioneering activities. His clear understanding, both of the power and also of the limitations of the new methods, did much to keep men's minds from jumping to hasty conclusions.

Huggins's early stellar investigations were made with spectroscopes attached to a small equatorial, with an aperture of only 8 inches. They were carried out with such conspicuous success that the Royal Society in 1869 decided to employ a large bequest made to them by Benjamin Oliveira in obtaining " a telescope of the highest power that is conveniently available for spectroscopy and its kindred inquiries. The instrument will, of course, be the property of the Society, and will be intrusted to such persons as, in their opinion, are the most likely to use it to the best advantage for the extension of this branch of science ; and in the first instance there can be but one opinion that the person so selected should be Mr. Huggins." The instrument was made by Sir Howard Grubb, and Stokes, who was then Secretary of the Royal Society, took great interest in the optical questions involved in the construction of the object glass. The installation of the new instrument provided Huggins with new opportunities, which he utilised to the full in the following decades.

Nasmyth's observations of detail on the surface of the sun, communicated to the *Memoirs of the Literary and Philosophical Society of Manchester* in 1862, and illustrated by plates which were reproduced at the end of Herschel's *Outlines of Astronomy* (8th ed., 1865), attracted wide attention ; and Dawes, Howlett, John Phillips, Stone, De la Rue, and Herschel joined in the lively discussions that followed. So much attention was paid to the description of " entities " that the essential feature of observation—namely, the fine-grained inequalities in the luminosity of the sun's surface—was in danger of escaping notice. But the discussions served the purpose of calling attention to the study of the sun's surface and in particular to the fine detail to be seen in and around sun-spots. Sir John Herschel introduced Howlett's beautiful delineations of sun-spots to the notice of the Society. Howlett continued his drawings for about thirty years, and they were presented by him to the Society, and form a notable collection of special interest on account of their bearing on the tenability of Alexander Wilson's idea (1773) of sun-spots as depressions.

Lockyer's attack on the problems of solar physics, which began in this decade with the help of the spectroscope, paved the way to great advances, leading to the foundation of the Solar Physics Committee and the Observatory at South Kensington in 1879, and doubtless also in large measure to the establishment of large observatories abroad, like that at Potsdam in 1874 and at Meudon in 1876.

Warren De la Rue [1815–89], who was elected President in 1864, was the eldest son of Thomas De la Rue, the founder of the eminent firm of manufacturing stationers of Bunhill Row. He became a good engineer without having received any special training, and his shrewd inventive faculty proved of great value to the firm. We owe so much to De la Rue for his early development of the application of photography to astronomy, that it is of interest to trace the history of his work. His earliest scientific papers relate to chemical and physical researches, doubtless suggested from time to time by the requirements of his firm. James Nasmyth [1808–90] has recorded, in his attractive reminiscences edited by Smiles in 1883, how De la Rue had visited him in 1840 to consult about mechanical appliances for a new process for the production of white lead, and had then seen the process of casting some disks of speculum metal for reflecting telescopes.

" I was then busy with the casting of my 13-inch speculum. He watched my proceedings with earnest interest and most careful attention. He told me many years after, that it was the sight of my special process of casting a sound speculum that in a manner caused him to turn his thoughts to practical astronomy, a subject in which he has exhibited such noble devotion as well as masterly skill."

Nasmyth cast a disk 13 inches in diameter for him, and out of it De la Rue constructed his celebrated reflector, which he set up in his garden in Canonbury, and later at Cranford. It is clear that before 1852 December, De la Rue must have worked several such mirrors. For in his notes on the figuring of specula (*M.N.*, **13**, 44), in which he describes his polishing machine, he says : . . . " I usually succeed in producing thirteen-inch mirrors, which define the planets . . . in a manner rarely equalled and never surpassed by any of the refractors which I have yet had an opportunity of looking through. I am, however, free to confess that they [the refractors] defined a fixed star much more satisfactorily than my best mirrors." He acknowledges his obligations both to

Lassell for his ready communication of his methods of manipulation, and to Nasmyth for many most valuable hints in fine grinding and polishing ; and he describes his own figurer and polisher.

He set up his best mirror, of aperture 13 inches and focal length 10 feet, equatorially mounted but without clockwork, in the garden of his house in Canonbury, and with it he studied the planets and made admirable drawings of Saturn, and Mars, and Jupiter, which were published from time to time. He was elected a Fellow of the Society on 1851 March 14. This was the year in which Archer applied collodion to photography, and suggested the use of pyrogallic acid for developing the latent image.

De la Rue was quick to seize upon the newly invented collodion plates, and in the autumn of 1852 he made " some successful positive lunar photographs in from ten to thirty seconds on a collodion film, by means of an equatorially mounted reflecting telescope of 13 inches aperture and 10 feet focal length, made in my workshop, the optical portion with my own hands ; and I believe I was the first to use the then recently discovered collodion in celestial photography."

It was not till 1857 that De la Rue fitted suitable driving clockwork to his reflector, when he moved it to his new residence at Cranford, about twelve miles west of London. There he raised his telescope on a pier 15 feet high, and arranged a photographic laboratory beneath the floor of the dome. From that year onwards we find in our *Monthly Notices* brief records of the work done at his observatory ; and from them we learn that De la Rue was constantly striving to surpass his best. He obtained new and more perfect photographs of the moon, " which will be of great value in forming selenographic charts and in showing correctly the extent and direction of the moon's libration."

De la Rue's admirable work in establishing solar photography might possibly have come out of his own pioneering instincts without any direct external impulse. But, as a matter of fact, the actual development of the idea and its accomplishment arose out of the discovery by Schwabe, of the periodicity of sun-spots, and the subsequent discovery of the identity of the period with that of magnetic disturbances.

The Kew Observatory building had been erected by King George III. for observing the transit of Venus in 1769, and after being maintained as the King's Observatory for seventy-one years it had passed, in 1842, into the management of the British Association for the Advancement of Science, and was used for research in meteorology and terrestrial magnetism and for the testing of scientific instruments. In 1856 the new photographic telescope, the construction and regular use of which for recording the state of the surface of the sun had been repeatedly advocated by Sir John Herschel, was installed in the dome on the top of the Observatory building under the direction of De la Rue.

A couple of years sufficed to overcome most of the difficulties of the novel work, though the illness and death of John Welsh, Superintendent of the Observatory, delayed progress. Welsh was succeeded by Balfour Stewart, and then began that scientific partnership between De la Rue and Stewart which is made memorable by their joint communications published in the *Philosophical Transactions* of 1869 and 1870.

In 1860 it was decided that the photoheliograph should be put at De la Rue's disposal for the total eclipse of the sun of 1860 July 18 ; and the occasion was made memorable by the successful photographic operations ; for they served to prove conclusively that the brilliant extrusions seen round the dark limb of the moon were indeed prominences connected with the sun, and, like the sun itself, subject to gradual eclipse by the moving moon. When De la Rue had completed the reduction of the observations with the help of a special micrometer devised by himself, he found himself impelled, by reason of the heavy weight of magnetic and meteorological work undertaken at Kew, to undertake the daily solar observations at Cranford. The optical tube of the photoheliograph was removed thither at the beginning of 1862, and was attached to the equatorial, where it remained at work until 1863 February. The instrument was then re-erected in the dome at Kew, and continued there in active operation for ten years more. In 1873 February it was moved to the Royal Observatory, Greenwich, and the fine series of solar records has been continued there with frequent improvement in the instrumental equipment and in the contributions of photographic records sent from observing stations in various parts of the globe.

It is interesting also to note the part played by Carrington's work in furthering the initial aims of the research. In the preface of his " *Observations of Sun-spots* " (1863), Carrington recounts how, when he set up his observatory at Red Hill, in the summer of 1852, for meridian observations of circumpolar stars, he was led to examine a series of drawings of the sun's disk in the possession of the Society. The discovery of the similarity in phase of the periodic recurrence of sun-spots and of magnetic disturbances had served to enhance the value of Schwabe's observations of the spots. Carrington had hopes of deriving, from observations extending over eleven years, the means of tracing system in the distribution and possible movements of spots and of detecting the true period of rotation. Even after Sir John Herschel, in 1854, had recommended photographic methods, Carrington with practical insight calculated on the probable slowness with which this method would be brought actually into application. Thus it came about that he secured his fine series of visual observations between 1853 November 9 and 1861 March 31, and that the series of drawings made by him on a uniform scale, such that the disc of the sun was a foot in diameter, were available for the measurement of sun-spot areas before the Kew photographs were systematically ready in 1862 February. Carrington's rotation period and his determination of the sun's axis were utilised in all the later reductions.

We can well understand that the fact that Carrington and De la Rue were Fellow-Secretaries of the Society during the five years 1857–62 contributed in no slight degree to the successful outcome of their labours. De la Rue gave two addresses during his Presidency, on the presenting the Gold Medal of the Society, first in 1865 to George Phillips Bond, for his work on the Comet of Donati, and then in 1866 to John Couch Adams, for his contributions to the development of Lunar Theory.

G. P. Bond [1825–65] succeeded to the Directorship of the Harvard College Observatory after the death of his father, William Cranch Bond, in 1859 January. He had been an assistant at the Observatory for many years during his father's directorship. The Observatory had been founded in 1839. G. P. Bond had made independent discovery of no less than eleven comets. It is curious that he never mentions in his Memoir on Donati's Comet, which forms volume **3** of the *Harvard Observations*, and was published in 1862, that the discovery of the comet was made by Donati, nor does he explicitly state the date of discovery, 1858 June 2. The whole period of visibility of that wonderful comet extended from 1858 June 2 to 1859 March 4, an interval of 275 days : it was visible to the naked eye for 112 days, and the tail was visible for 177 days. Bond's Memoir deals with the tails, nucleus, and envelopes of the comet, and is illustrated by more than fifty plates, which exhibit in an admirable manner the changes that occurred from time to time in the wonderful phenomena presented by the comet.

The Council's Annual Report in 1859 gave a note of nine pages about the comet. It is not easy to trace the authorship of the early Council Notes. Carrington and De la Rue were Secretaries at the time, and it is not improbable that this note was prepared by one or both of them. It is of very unusual length for the notes contributed in those days, and gave a valuable summary of the phenomena. De la Rue's address gives us some idea of the attention he had himself given to the comet, and it seems not unnatural to attribute the note to him.

The announcement of Bond's death on 1865 February 17, seven days after the delivery of De la Rue's address, was made in the same number of the *Monthly Notices* that contained the address. At a later meeting the President was able to assure the Society that Bond had known of the award before his death, and greatly appreciated the recognition of his labours. Among his other achievements, Bond had discovered the crape ring of Saturn and also Hyperion, the eighth satellite of Saturn. Independent discovery of these had been made respectively by "eagle-eyed" Dawes and by Lassell.

De la Rue's second address was given in 1866 February on the award of the medal to Adams. It was a very able address, and in the preparation of it, as Dr. Glaisher has recorded in his obituary notice of Adams (*M.N.*, **53**, 199), he had the invaluable assistance of Delaunay, and gave an excellent history of the problem of the secular acceleration of the moon.

Adams [1819-92] was President of the Society in 1851 and 1852, and in the latter year he communicated to our Society new tables of the moon's parallax. In 1853 he communicated to the Royal Society his celebrated paper (of ten pages) on the secular acceleration of the moon's mean motion. It was for these two pieces of work in the development of Lunar Theory that the award of the medal was made in 1866.

In the interval of fourteen years a great controversy raged. In 1853, Laplace's discovery in 1787 that the secular variation of the eccentricity of the earth's orbit produced secular terms in the moon's motion was still regarded as having set the question of the moon's acceleration at rest. Adams showed that the true value of the acceleration requires the insertion of additional terms into the equations, and the result of such insertion is to decrease the then accepted value of the acceleration. Plana remonstrated, Delaunay intervened. Pontécoulant attacked, Hansen calculated. Le Verrier inclined towards an incorrect theory because observation supported it. Lubbock, Donkin, and Cayley joined in the calculations and discussions. " The whole controversy forms a very extraordinary episode in the history of physical astronomy ; the indifference with which Adams's Memoir of 1853 was at first received, in spite of the interest and importance of the subject, being followed by the violent controversy which resulted in so many independent investigations by which Adams's result was confirmed " (Glaisher, *M.N.*, **53**, 198). Delaunay's account of the whole discussion appeared in the supplement of the *Connaissance des Temps* for 1864. He says, " L'apparition du mémoire de M. Adams a été un veritable événement ; c'était toute une révolution qu'il opérait dans cette partie de l'astronomie

théorique." Delaunay confirmed Adams's value of the acceleration attributable to secular variation of the eccentricity of the
earth's orbit, and assigned the outstanding discrepancy between
observation and calculation to the existence of another source of
variation, namely, the secular lengthening of the day by the
action of tidal friction.

Adams had the great satisfaction not only of the recognition
of his own labours, but also of doing honour to Delaunay, when in
1870 he delivered the address on presenting the Gold Medal of
the Society to him.

The Rev. Charles Pritchard [1808–93] was elected President
of our Society in 1866. He had entered St. John's College, Cambridge, in 1826, and was fourth Wrangler in the Mathematical
Tripos in 1830. Two years later he was elected to a Fellowship
in his college, and in 1834 he became the first headmaster of
Clapham Grammar School, founded as a new venture by the
leading men in Clapham. The new school, with its eager enthusiastic master, attracted pupils from all parts of the kingdom,
and men distinguished in science and the liberal professions sent
their sons to him to benefit by the breadth and originality of his
teaching. In 1862 he retired from the headmastership. He had
joined our Society in 1849, was elected on the Council in 1856,
and served as Secretary from 1862 to 1866. His election to the
Savilian Professorship of Astronomy at Oxford, in succession to
Professor Donkin, in 1869, was the beginning of a new opportunity,
in which he showed untiring zeal and energy.

Pritchard delivered two addresses during his Presidency, on
presenting the Gold Medal of the Society, first in 1867 to William
Huggins [1824–1910], and William Allen Miller [1817–70] conjointly for their researches in Astronomical Physics, and then in
1868 to Le Verrier for his Planetary Tables.

The award of the Gold Medal to two persons conjointly in 1867
involved a suspension of certain of the Bye-laws. The ordinary
meeting of the Society in January was followed by a Special
General Meeting, at which a resolution was passed empowering
the Council " to award the medal to two gentlemen who have been
engaged conjointly in a work of astronomical importance." The
Council reassembled after the Special General Meeting, and proceeding to the ballot, they awarded the medal to Huggins and Miller
conjointly. At the February Meeting the Council resolved that
the medal be engraved in duplicate with the names of the two
recipients.

Little did the President or anyone else at the time think of
the revolution that was to come in the next quarter century.

But the first-fruits gathered in those early years were remarkable enough, showing as they did that the stars differing from one another in the kinds of matter of which their spectra could give evidence are constructed upon the same model as the sun, and are composed of matter identical, at least in part, with the materials of our own system. Huggins's work on the nebulæ roused even greater interest than his studies of the stellar spectra. Lord Rosse had inferred from observations made with his 6-foot reflector at Parsonstown, that many nebulæ which had not been resolved into starry clouds with less powerful instruments could be seen in his giant telescope as clusters of minute stars. The year 1864 saw Huggins's discovery of the gaseous nature of eight planetary nebulæ, proving that they could no longer be regarded as aggregations of suns after the order to which our sun and the fixed stars belong, and that we must regard them as enormous masses of luminous gas or vapour. The first steps in the solution of part of the mystery of a comet as well as that of a new star were made by Huggins in the years of this decade.

Some confusion still exists between the work of William Allen Miller and William Hallowes Miller [1801-80]. The former was Assistant Lecturer to Professor Daniell at King's College, London, and became Professor of Chemistry in succession to Daniell at his death in 1845. William Hallowes Miller was Professor of Mineralogy at Cambridge (1832-80). It was he, who in 1833 made experiments conjointly with Daniell, on the discontinuous absorption spectra of iodine and bromine. It was he also who verified the coincidence of the bright sodium lines with the dark D lines in the solar spectrum some years before Kirchhoff's classical work in 1859.

In the correspondence which passed between Thomson and Stokes five or six years before Kirchhoff's celebrated Memoirs, and which was found by Sir Joseph Larmor in arranging the scientific correspondence of Stokes (*Collected Papers*, **4**, 367), references to the work of Miller occur in several places ; and the Miller there named was William Hallowes Miller.

The meeting of the Society on 1866 December 14 was a specially interesting one, for at it were received accounts of the observations of the Leonids. A. S. Herschel gave the position of the radiant determined by fifteen observers. Airy told of 8500 meteors having been counted at Greenwich.

In anticipation of a display of meteors, such as had been observed by Humboldt at Cumana, near the northern coast of South America, in 1799, and by many observers in 1833, considerable attention was given to the observation of meteor tracks, especially in the years of this decade. Humboldt had failed to recognise in 1799 that the meteors radiated from a point fixed in relation to the stars. It was Olmsted, of Yale University, who first in 1834 recognised the significance of this point as indicating the direction of the meteors in their approach to the earth, and he regarded them as a form of comet describing an elliptical orbit with a period of about 182 days, and meeting the earth near aphelion. Erman, of Berlin, discussing in 1839 the similar problem presented by the August meteors (Perseids), found it necessary to assume that the meteors in that case formed a continuous stream along their orbit.

Olbers, in 1839, was led to predict a fine display of Leonids in 1867 November. But fortunately H. A. Newton, Professor in Yale University, published in 1864 his well-known discussion of ancient records of November meteors, dealing with thirteen showers since A.D. 902, and indicating the existence of a cycle of 33·25 years. Considering the phenomena to be caused by a ring of meteoroids revolving round the sun, he showed that in one year the meteoroids must describe either $2 \pm \frac{1}{33\cdot25}$ or $1 \pm \frac{1}{33\cdot25}$, or $\frac{1}{33\cdot25}$ revolutions. He further pointed out that the longitude of the node of the orbit is gradually increasing, and that its observed motion would afford a method for deciding which of the five periods is the correct one, if only the perturbations by the various planets were calculated. He predicted a fine display of meteors for 1866 November, a year earlier than Olbers's date. When Newton's prediction was verified, the problem became a very attractive one. It was made all the more attractive by Schiaparelli's discovery that the cosmical orbit of the Perseid meteors coincided closely with the orbit of the retrograde comet which was discovered by Swift in 1862, and which reached perihelion on August 22 of that year.

The spring of 1867 is made memorable by a display of striking Memoirs following one another with almost meteoric rapidity. In January Le Verrier published his Memoir showing that a swarm of meteors with a period of 33·25 years would intersect the orbit of Uranus, but from its inclined position indicated by the radiant's latitude 10° it would not intersect the paths of Saturn, Jupiter, or Mars. His calculations showed that in A.D. 126 there would have been a close approach of Uranus to such a swarm, and that that date might be the epoch of the capture of the swarm for the solar system by their diversion into a retrograde elliptic orbit of period 33·25 years.

In 1867 February, C. F. W. Peters and Oppolzer pointed out the close resemblance of Oppolzer's orbit for the comet discovered by Tempel in 1865 December, which reached perihelion on 1866 January 11, to Le Verrier's orbit of the meteors.

In the same month, Schiaparelli had from his own calculations of the meteoric orbit made the same comparison.

Yet again in February, Le Verrier, noting Peters's suggestion, had recalculated the meteoric orbit, utilising A. S. Herschel's determination of the radiant in 1866, and had found a better agreement with Oppolzer's orbit.

And in 1867 March, Adams had completed the calculations of the planetary perturbations, and had found that the observed variation of the node of the meteoric orbit could not be reconciled with the four shorter periods indicated by H. A. Newton, but was completely satisfied by the longest period. From a combination of five new determinations of the radiant with that derived from his own observations with an instrument specially devised by him, he deduced a definitive orbit still more closely in agreement with Oppolzer's orbit for Tempel's comet (*M.N.*, **27**, 247). And thus was established the close relation between comets and meteors.

Pritchard's second Presidential Address was delivered in 1868, on the occasion of the award of the Gold Medal to Le Verrier, for his theories and tables of the four planets, Mercury, Venus, Earth, Mars. It was with this work that Pritchard dealt in his address. For his later work on the theories of Jupiter, Saturn, Uranus, and Neptune, the medal was again awarded to Le Verrier in 1876, and on that occasion Adams delivered the address. The earlier work led Le Verrier to infer that there existed on the one hand in the neighbourhood of Mercury, and on the other hand in the neighbourhood of Mars, sensible quantities of matter, the action of which had not been taken into account. In the case of Mars, the mass of the earth itself was at fault ; it had been assumed too small, having been derived from too small a value of the solar parallax. " With respect to Mercury, a similar verification has not yet taken place, but the theory of the planet has been established with so much care, and the transits of the planet across the sun furnish such accurate observations as to leave no doubt of the reality of the phenomenon in question ; and the only way of accounting for it appears to be to suppose, with M. Le Verrier, the existence of several minute planets, or of a certain quantity of diffused matter circulating about the sun within the orbit of Mercury " (Adams, *M.N.*, **36**, 232).

Considerations like these, set forth by men like Le Verrier and Adams, even though half a century ago, still carry weight with those who hesitate to accept the astronomical evidence of the deflection of light in a gravitational field as a crucial verification of the truth of Einstein's theory ; to them the astronomical evidence is less strong than that to be derived from other physical phenomena.

Admiral Russell Henry Manners [1800–70] was elected President in 1868. He had retired from active service in 1849. He joined the Society in 1836 and was one of the Secretaries for ten years, 1848-57; he then became Foreign Secretary, and held that post for ten years, till his election as President.

Manners delivered the address in 1869 on presenting the Gold Medal of the Society to Stone, for his rediscussion of the Transit of Venus in 1769 and his other contributions to Astronomy. By reason of illness, Manners was unable to give the address on the award of the medal to Delaunay in 1870 February, and he died on May 9. As has been stated on an earlier page, Adams delivered the address on Delaunay's work.

E. J. Stone [1831–97] was educated at King's College, London, and in 1856 went to Queens' College, Cambridge. He graduated as fifth Wrangler in 1859 and became a Fellow of his College. In 1860 he became First Assistant at the Royal Observatory, and held the post for ten years, until he was appointed to the Cape Observatory in 1870 on the resignation of Sir Thomas Maclear. On Main's death in 1878, Stone was appointed Radcliffe Observer, and held the post until the day of his almost sudden death in 1897. He was one of the Secretaries of the Society from 1866 till 1871, Huggins being his fellow Secretary.

Stone's work on the solar parallax was the chief theme of Manners's address. Airy had in 1854 and 1857 taken steps to move astronomers to a realisation of the need to revise Encke's value $8''{\cdot}57$, derived from the observations of the 1761 and 1769 Transits of Venus, which had been held in general acceptance from 1824 up to the beginning of this decade. Hansen had also derived from his investigations of the moon's motion a conviction that Encke's value was too low (*M.N.*, **15**, 9), and had himself deduced a value $8''{\cdot}92$ (*M.N.*, **24**, 8). Again Le Verrier had found in 1858 from his researches that a value of $8''{\cdot}95$ was indicated. Yet again Foucault in 1854 had measured the velocity of light and found a value 185,287 miles/sec. (298,187 km./sec.), considerably less than the value deduced by Fizeau from his avowedly preliminary experiment in 1849 (194,663 miles/sec. or 313,274 km./sec). Observations of eclipses of Jupiter's satellites were taken as indicating a value of about 192,500 miles/sec. All these lines of research called for a revision of the value of the sun's distance from the earth. Hence Airy's eagerness for a worthy attack on the problem, and the readiness with which astronomers responded to the call. Stone's reductions of observations of Mars made in 1862 at Greenwich, the Cape, and Williamstown, Victoria, resulted in a value of the solar parallax 8"·94 (*Mem. R.A.S.*, **33,** 77). Similar reductions made by Winnecke, of observations made in 1862 at Pulkova and the Cape, resulted in a parallax of 8"·96. From a rediscussion of the observations of the Transit of Venus 1769, Stone, by a reasonable interpretation of some of the descriptions given by observers, deduced a solar parallax 8"·91 (*M.N.*, **28,** 255).

Thus it came about that Le Verrier's value of 8"·95 was introduced in the *Nautical Almanac* for 1870, and continued in use until the *Almanac* for 1882, when Newcomb's value 8"·85 was adopted.

Great preparations were made in 1868 for observing the total eclipse of the sun on August 17–18. Two expeditions were sent out from this country to India, one under Major J. T. Tennant, arranged by our Society, with the financial aid of the Indian and Imperial Governments : and the other under Lieut. John Herschel, R.E., arranged by the Royal Society, with financial aid from the Parliamentary grant annually placed at the disposal of that Society.

This eclipse is made memorable by the success of the observations which enabled both Tennant and Herschel to announce that the solar prominences exhibited bright lines in their spectra, indicating at any rate the presence of incandescent hydrogen and probably sodium and magnesium. Janssen's observations during the eclipse convinced him that he would be able to see the prominences with the help of his spectroscope in full sunlight, and he recounted how for two or three days after the eclipse he had been living in a veritable fairyland of new observations. Lockyer, working in London, with apparatus which had long before been designed for this particular research, but of which the completion had been delayed for several months in a busy optician's workshop, was able to make announcement of the success of his observations. It arrived at the Paris Academy only a few minutes before Janssen's report from India, and the two investigators share the honour of the discovery of the new method.

Another result of this eclipse was to bring about joint action between the Royal Society and our Society in making arrangements for the observations of the eclipse in 1870 December. This joint action was renewed from time to time, until in 1892 it ceased to be temporary by the appointment of a Joint Permanent Eclipse Committee of the two Societies.

In reviewing the impressions gained in reading through the records available for this chapter in the history of the Society, one is led to feel that it was a decade of great and wholesome activity. The heritage of large problems from the previous decade was one that called for great performance; and the part played by the men in responsible positions was worthy of the heritage. Airy and Adams were the most marked men, and the Royal Observatory stood out pre-eminent in contributing solutions of questions of high importance in the Science, whilst De la Rue, Carrington, Huggins, and Lockyer were breaking new ground that was to yield splendid harvest in later years.

In the accounts of the meetings one gathers something of the personality of the conspicuous men : Airy, a dominating figure, unbending and gruff; Adams, clear-minded, quiet, and helpful; Pritchard, lively and sympathetic; Lee, stately and courteous; Carrington, impetuous; Huggins, careful and judicial; and De la Rue, a man of order and energy, on cordial terms with everyone.

Of the growing prosperity of the Society there were many indubitable signs. In the middle of the decade the Council prefaced their Annual Report as follows (*M.N.*, **26,** 101) :—

> The Council cannot recollect any former occasion on which there has been better ground for congratulation to the Royal Astronomical Society than at the close of the past year. Looking backwards ten years, they find the number of the contributing members has increased by nearly thirty per cent. The attendance at the Evening Meetings has more than doubled, and the funded property of the Society, during the nine years' tenure of office by the present Treasurer, has increased by upwards of £2700 stock. Applications for the supply of the *Monthly Notices* of our proceedings continue to be made from every quarter of the globe; and several of the numerous private Observatories scattered throughout the country are showing signs of increasing vitality by the production of fresh and valuable results.

The situation in Astronomy at the end of this decade was well summed up by Stokes in his Presidential Address to the British Association at Exeter in 1869. After referring to advances made in dynamical astronomy, he spoke as follows :—

> After these brilliant achievements, some may perhaps have been tempted to imagine that the field of astronomical research must have been well-nigh exhausted. Small perturbations, hitherto overlooked, might be determined, and astronomical tables thereby rendered still more exact. New asteroids might be discovered by the telescope. More accurate values of the constants with which we have to deal might be obtained. But no essential novelty of principle was to be looked for in the department of astronomy; for such we must go to younger and less mature branches of science.
>
> Researches which have been carried on within the last few years, even the progress which has been made within the last

twelve months, shows how shortsighted such an anticipation would have been ; what an unexpected flood of light may sometimes be thrown over one science by its union with another.

Then follow references to the work of Bradley, Huggins, and Miller.

> The determination of radial proper motion in this way is still in its infancy. It is worthy of note that, unlike the detection of transversal proper motion by change of angular position, it is equally applicable to stars at all distances, provided they are bright enough to render the observations possible. It is conceivable that the results of these observations may one day lead to a determination of the motion of the solar system in space which is more trustworthy than that which has been deduced from changes of position, as being founded on a broader induction and not confined to conclusions derived from the stars in our neighbourhood.

Herewith we close this chapter, feeling assured that with all its inadequacy it can hardly have failed to show that the Society was handing on a heritage calling for great endeavour.

CHAPTER VI

THE DECADE 1870–1880. (By H. P. Hollis)

1. 1870–1873

In 1870 February fifty years had elapsed since the foundation of the Society, and this Jubilee was recognised by an attempt to secure a specially large attendance at the dinner, which it was then customary to have on the day of the Annual General Meeting. A circular-letter was issued to Fellows inviting them to be present " to make the dinner an occasion of commemorating the foundation of the Society fifty years since." * M. Delaunay, the recipient of the Gold Medal, was at the dinner. The address delivered at the meeting on presentation of the medal was written by Professor J. C. Adams, Vice-President, and was read by him, as the President, Admiral Manners, was absent through illness.†

At this February meeting, Mr. Lassell was elected President for the ensuing year, and Mr. Huggins and Mr. Stone retained office as Secretaries. The partnership did not continue long, because Mr. Stone was appointed to the post of H.M. Astronomer at the Cape, in June, on the resignation of Sir Thomas Maclear, and the duties of the Secretaryship were undertaken temporarily by Prof. Pritchard, Mr. Burr, or Mr. Dunkin. At the meeting of Council in November, Mr. Dunkin, of the Royal Observatory, Greenwich, and Mr. Huggins, were nominated Secretaries in the list of officers to be submitted at the February meeting.

Mr. Proctor's name had been proposed for the Secretaryship, but he declined the office as he was unable to give sufficient time to the duties. As he took an active part in the life of the Society in the years now to be written of, some extracts from his biography may not be out of place. Mr. Proctor took his degree as twenty-third Wrangler in 1860, and without following any settled profession, occupied himself in writing occasionally on astronomical subjects. His first book, *Saturn and its System*, which took four years in preparation, was published in 1865. In 1866 he suffered

---

* " The dinner will be at Willis' Rooms, at half-past 5. Price, including wine, 20s."

† He died in the following May.

a severe pecuniary loss through the failure of a New Zealand bank, and was practically compelled to earn his living by authorship. He had joined the Society in 1866 and began his long series of contributions in the *Monthly Notices* by papers in 1867 and 1868 on the rotation of Mars. The coming transits of Venus in 1874 and 1882 were now engaging attention, and in 1868 December, Airy presented a paper " On the Preparatory Arrangements which will be necessary for efficient Observation of the Transits," in which he pointed out that the method of observation known as Halley's, failed for the transit of 1874, and though it was suitable for that of 1882 there was a difficulty in finding a suitable southern station ; and that it would be advisable to plan the observations of the first of the pair according to Delisle's method. Mr. Proctor communicated a paper to the Society in 1869 March, showing that Halley's method was quite suitable for the 1874 transit, a view taken by M. J. Puiseux in a contribution to the *Comptes Rendus* of 1869 February, and he followed this with a more complete and detailed paper read at the meeting in May, which will be found in the *Monthly Notices* of 1869 June. Mr. Stone took up the discussion on the part of the Astronomer Royal. The feeling that existed with regard to this matter appears from the following extract from a letter written by Mr. Proctor to the Astronomer Royal on 1869 May 15, the day after the meeting of the Society :—

> The high respect and esteem in which the scientific world holds the name of the Astronomer Royal for England is shared in by no one more fully than by myself. But I should consider that no greater discourtesy could be shown to that name than by an attempt to modify statements of scientific fact in presence of it. Mr. Stone praised such a course on M. Puiseux's part as a piece of courtesy ; but M. Puiseux must in justice be acquitted of so serious an offence against scientific morality and of what would have been a gross rudeness towards yourself.
>
> I think I stated the simple truth when I said last night that no man living has so earnest a desire to see the coming transits properly utilised as yourself : and I conceive that in pointing to a certain application of Halley's method in 1874 as the most powerful mode of determining the sun's distance available until the twenty-first century, I was fulfilling what no one would admit more readily than yourself to be a duty.

The subject of the Transit of Venus and this discussion will be treated in more detail later, as the incidents are being related to some extent in chronological order.

At the meetings in the year 1870 the subject of observation of the total solar eclipse of December 22 was repeatedly brought before the Society. The central line of this eclipse passed over the south coast of Spain, Sicily, and the north coast of Africa, and it was proposed to repeat the experience of the eclipse of 1860, when the Admiralty provided a ship to convey an expedition to a convenient spot for observation.   At the meeting in March, Mr. G. F. Chambers,* asked whether the Government proposed to help astronomers in going out to Spain to observe the eclipse. He was told that the Government did not *propose* to do anything of the kind, but that the Council were prepared to lay before the authorities a statement of the required observations, and to urge the necessity for some assistance.†   The Council had on that day resolved itself into a Committee to consider the preparations necessary, and in the following month a Committee of the Society united itself with a Committee of the Royal Society appointed for the same purpose, Professor Stokes being Secretary of the joint body.   The Society voted the sum of £250 towards the expenses, to which the Royal Society added an equal sum.   The Astronomer Royal, as spokesman for the two Societies, asked the Admiralty to supply two ships, one to take observers to Spain, the other to Syracuse in Sicily; but the application was not acceded to, and a further application was therefore made to the Treasury.   It was not until the day of the November meeting (the 11th) that the Committee were able to count on any help from the Government, and it appears that this help was given, after a previous definite refusal, by reason of the intervention of Mr. Lockyer who explained the necessity to the officials concerned.   The use of H.M.S. *Urgent* was granted to carry observers to Spain, and the sum of £2000 was contributed by the Treasury towards the expenses.   An organising Committee was appointed, of which Mr. Norman Lockyer was Secretary, and Mr. Ranyard, Assistant Secretary.

The first-named has already been mentioned.   He was at this time a clerk in the War Office, but was already famous in the astronomical world for his solar spectroscopic researches, and specially for his suggestion, made in 1866, and his discovery in 1868 of the method of observing prominences at times other than when the sun is eclipsed, the credit for which is shared with M. Janssen.   Mr. A. C. Ranyard, who was a prominent figure in the affairs of the Society later on, had joined the Society in 1863 at the age of eighteen, before proceeding to Cambridge, and at this time was reading for the Bar.

* An amateur astronomer who joined the Society in 1864, and the author of a well-known book, *Handbook of Descriptive Astronomy*.

† A circular (printed in the *Monthly Notices for* 1870 April) was sent by direction of the Council to all Fellows of the Society, inviting those who were willing to take part in the eclipse observations to send their names to the Secretaries.   It appears that fifty or sixty volunteered in response.

On December 6 the *Urgent* left Portsmouth, taking three parties, made up chiefly of Fellows of the Society under the leadership of the Rev. S. Perry, Captain Parsons, R.E., and Mr. Huggins, bound respectively for Cadiz, Gibraltar, and Oran, in Algeria, and with them were Professor and Mrs. S. Newcomb.   Another expedition under the leadership of Mr. Lockyer, comprising amongst others, Professor Roscoe, Mr. Darwin, Mr. Vignoles, and Mr. Ranyard, went by overland route to Naples, and left that port to cross to Catania (Sicily) in H.M.S. *Psyche*.   Unfortunately the vessel struck on a rock near Catania, but all hands, and the instruments, were saved without injury.   Lord Lindsay, who was not then a Fellow of the Society, took an observing party at his own expense to Cadiz. On the day of the eclipse the sky was more or less obscured by cloud at all the stations.   At Cadiz and at Syracuse successful photographs of the corona were obtained, as well as some spectrum and polarisation observations, but at Oran nothing was seen of the eclipse at totality.   The photographs of the corona taken at Syracuse by Mr. Brothers with a rapid rectilinear photographic lens, showed great extensions and were considered specially successful.

As indication of the state of knowledge of the sun's surroundings at the time, it may be remarked that at the meeting in 1870 June a paper by Mr. Seabroke " On the determination whether the Corona is a Terrestrial or Solar Phenomenon," led to a discussion on this fundamental point in Solar Astronomy.   Mr. Lockyer's " theory of a terrestrial origin of the corona " was spoken of, the reference probably being to an article by him in the first number of *Nature*, in which he said, " Since that time I confess the conviction, that the corona is nothing else than an effect, due to the passage of sunlight through our own atmosphere near the moon's place, has been growing stronger and stronger."   Dr. Gould, the American astronomer, who was at the meeting, spoke of his observations during the eclipse of 1869 August 7, and said that he thought the symmetry of the corona about the sun's axis of rotation pointed to the fact that it was of solar origin, and that the trapezoidal corona might be nothing more than the chromosphere seen under unusually favourable circumstances, but he was inclined to think that the light outside that four-cornered corona which appeared to shift in position was an effect of our atmosphere.

The Society as a body took a less active part in the arrangements for the observation of the solar eclipse of 1871 December 12, on which occasion the line of totality crossed India, Ceylon, and Australia.   The subject was brought before the Council at their meeting in June, when it was at first suggested that the Indian arrangements should be left in the hands of Mr. Pogson,

Director of the Madras Observatory, the Astronomer Royal consenting to act as an intermediary.   It was, however, proposed by Mr. Lockyer, Mr. De la Rue seconding, that the Imperial and Indian Governments should be asked to provide facilities for a few competent observers, who might volunteer, to proceed to India and Ceylon free of expense to make observations at those places. A Committee was appointed to consider this proposal, and at a meeting on June 28 it was resolved that the observations should be limited to a complete examination of the corona, and that Mr. Huggins and Mr. Lockyer should be asked whether they would go to India to undertake this.   At a subsequent meeting of the Committee on July 7, at which Mr. Lockyer was present, it was reported that Mr. Huggins could not go, but Mr. Lockyer expressed his willingness to do so if he could get leave from his duties.   He also mentioned that he knew that the officers of the Royal Society were prepared to join the Royal Astronomical Society Committee in applying to the Treasury for a grant, and gave the outlines of a scheme of observation that he had prepared, which included spectroscopic as well as photographic observations of the corona. The Committee resolved that they had no power to form part of a joint Committee, and adhered to their resolution of June 28, that only a complete examination of the corona, not comprising spectroscopic work, should be attempted.   To this Mr. Lockyer would not agree, and he declined to go.   Under these circumstances it was decided to proceed no further in the matter, and the organisation of observation of the eclipse passed into the hands of the British Association.   At the Annual Meeting of the Association, which was held at Edinburgh in that year, a Committee was formed, Lassell, De la Rue, Airy, Stokes, and Lockyer being the active members, to take in hand matters relating to the eclipse.   A grant of £2000 was obtained from the Government, and eventually an observing party went to Ceylon under the leadership of Mr. Lockyer, and met with complete success.   Arrangements for observation of the eclipse in India were put in the hands of Major J. F. Tennant by the Indian Government, who provided the necessary resources. This expedition, and those sent out by other Governments to various stations on the line of totality, were equally successful, but the observers in Australia were not favoured with good weather conditions.

It seems appropriate to mention here the origin of the volume of the Society's *Memoirs* (**41**) known generally as the Eclipse Volume.   It was the outcome of a suggestion by Airy, that the results obtained by the observers of the eclipses of 1860 and 1870, who were subsidised by the Government, should be published at public expense.   The Treasury refused to grant funds for the purpose, and Airy consequently wrote a letter to the Council on 1872 January 12 informing them of this refusal and asking for financial help from the Society for his scheme. The Council resolved that the account should be published as a *Memoir* of the Society. This was eventually done, the preparation being confided to Mr. Ranyard, who at first worked under the direction of Sir George Airy. Later on, the entire work devolved upon him, and the scheme of the volume was enlarged. Consent was obtained from the British Association to include the results of the eclipse of 1871, and those of earlier eclipses were added, so that finally it became a record of all observed eclipses up to 1878. The volume was completed after a delay, which caused occasional remark, and was published in 1879.

No medal was awarded in the year 1871, and the reasons for this are of interest. At the Council Meeting in 1870 November, it was proposed by Mr. Dunkin, Mr. De la Rue seconding, that Mr. Lockyer should receive the medal for his researches in Solar Physics. Other recipients were proposed, and lastly it was proposed by Mr. Browning, Professor Pritchard seconding, that Mr. Lockyer and Dr. Frankland should receive the medal for their joint researches in Solar Physics.

At the Council Meeting in 1870 December the ballot was taken and Mr. Lockyer's was the name chosen to be submitted for confirmation at the meeting in January. But at that meeting, after considerable discussion, the choice was not confirmed. At the Annual General Meeting in 1871 February the President stated, as may be read in the account of the meeting in the *Monthly Notices*, that the non-award was due to the fact that by the Bye-laws the Council had not the power to bestow a joint medal, and that the Council had " found it impossible to select any individual so pre-eminently distinguished by his own independent researches that they could recommend the Society to bestow its award upon him, without danger of doing injustice to others." But it is not difficult to infer from the discussions at the Council Meetings, and from subsequent events, that there was a feeling inimical to Mr. Lockyer.

At a Special General Meeting of the Society held on 1871 June 9, a new Bye-law was proposed by the Council and passed, to the effect that " where two or more persons have been jointly concerned in the production of any scientific treatise, or the carrying out of any research work or discovery, or have been the simultaneous but independent authors of any such treatise, work, research, or discovery," the Council have power to award a medal to each of such joint authors. It will be remembered that the medal had been awarded to Mr. Huggins and Professor Miller jointly in 1867.

As to the award of the medal in 1872, at the meeting of the

Council in November 1871, the names of Mr. Lockyer and Dr. Frankland were proposed as recipients by Mr. De la Rue, and seconded by Mr. Browning, for their joint researches in Solar Physics and the Spectra of Gaseous Bodies.   The names of other astronomers were proposed, and of these Professor Schiaparelli was selected at the December Meeting and received the medal at the Annual General Meeting in 1872 February, for his Researches on the Connection between the Orbits of Comets and Meteors.

Beyond the communications relating to the arrangements for the eclipse of 1870 and the reports of the results, there was rather a scarcity of papers read before the Society in the years 1870 and 1871.   Mr. Proctor was the largest contributor, and there are more than twenty papers by him in these two years.

The star Eta Argus, and alleged changes in the nebula surrounding it, formed the subject of several communications in 1871.   A noteworthy paper by Professor Alexander Herschel will be found in the *Monthly Notices* for June of that year, which expounded an idea conceived by his brother, Captain Herschel, for the automatic registration of transits, and gives his own (Professor A. Herschel's) plan for carrying this out mechanically. The method of this apparatus is in effect precisely that of the type of registering transit micrometer brought into use twenty years later, in which the wire is moved by mechanical means, with the useful addition that a means was provided for the observer to suppress the record of a contact if he were not satisfied that the coincidence of the wire and star was perfect.

The Society lost by death three distinguished Fellows in the year 1871 : Sir John Herschel, Mr. Babbage, and Professor De Morgan, the two first-named were the last survivors of those who met at the Freemasons' Tavern in 1820 January to consider the expediency of establishing an Astronomical Society.

At the Annual General Meeting in 1872 February, Professor Cayley was elected President, Mr. Huggins retired from the Secretaryship, and Mr. Proctor was chosen to succeed him.

The following session, March–June, was remarkable for a discussion, or series of discussions, in the Council.   The subject found its way later into the public press under the heading of Government Aid to Science, or the Endowment of Research, and though action was not taken at this time, the proposal made to the Society in this session may be considered to have resulted in the establishment of the Observatory at South Kensington, which later on played such a large part in the development of Astrophysics.

The matter was initiated by Lieut.-Colonel Strange, the Foreign Secretary, who had been a distinguished officer of the Indian Trigonometrical Survey, and had retired from the army in 1861.

He was then appointed Inspector of Scientific Instruments for India, and in that capacity was the first Superintendent of the depot at Lambeth, where an Observatory was erected for him, in which he did much good work in designing and testing apparatus. He joined the Society in 1861, served on the Council almost continuously from 1863, and was chosen Foreign Secretary in 1870. Many papers on instrumental subjects will be found in the indexes under his name.

At the meeting at Norwich in 1868, the British Association, at the instance of Colonel Strange, took into consideration the subject of Government Aid for Science. This led to little result, but in April 1872, Colonel Strange, claiming to follow the lead of Sir George Airy, who a few months earlier had proposed that an observatory should be established solely for the observation of the phenomena of Jupiter's satellites, read a paper before our Society on " The Insufficiency of Existing National Observatories." In this he asserted that permanent national provision for the cultivation of the Physics of Astronomy was urgently needed. At the meeting of the Council in May he proposed that " the President be authorised, on behalf of the Fellows and Council of the Royal Astronomical Society, to bring before the Royal Commission on Scientific Instruction and the Advancement of Science, now sitting, the desirability of providing for the cultivation of the Physics of Astronomy." The Commission referred to was under the Chairmanship of the Duke of Devonshire, and had been in existence for two or three years. It had been established, it is said, largely owing to Colonel Strange's persistent advocacy of the necessity for Government aid for the promotion of scientific research. Mr. Lockyer was Secretary of this Commission. It had at the time practically completed the first part of its work, which was concerned with scientific instruction. After discussion by the Council in May, consideration of Colonel Strange's proposal was adjourned to the June meeting, and then to a Special Meeting of the Council on June 21, and again to another on June 28. At this last meeting the following motions proposed by Dr. Huggins were taken into consideration :—

1. That the President be authorised, on behalf of the Council and Fellows of the Royal Astronomical Society, to bring before the Royal Commissioners on Scientific Instruction and on the Advancement of Science, now sitting, the importance of further aid being afforded to the cultivation of the Physics of Astronomy.

2. The Council think such aid would be most effectually given by increased assistance where needed to existing Public Observatories in the direction recommended by the heads of those observatories, especially that at the Cape of Good Hope, and by the estab-

lishment of a new Observatory on the Highlands of India, or in some other part of the British dominions where the climate is favourable for the use of large instruments.

3. The Council do not recommend the establishment of an independent Government Observatory for the cultivation of Astronomical Physics in England, especially as they have been informed that the Board of Visitors of the Royal Observatory at their recent meeting recommended the taking of Photographic and Spectroscopic records of the Sun at that Observatory.

It was proposed in amendment by Colonel Strange, seconded by Mr. De la Rue, that the following be substituted for these clauses :—

The following seems to the Council to be the provision now requisite :—

1. An Observatory, with a laboratory and workshop of moderate extent attached to it, to be established in England for the above researches.

2. A certain number of Branch observatories, to be established in carefully-selected positions in British territory, in communication with the Central Observatory in England, for the purpose of—first, giving to Photographic Solar Registry that continuity which experience has already proved to be necessary ; and, *secondly*, to investigate the effect of the Earth's Atmosphere on Physico-Astronomical Researches in different geographical regions, and at different altitudes. In these purposes India and the Colonies offer peculiar advantages.

The amendment did not receive approval, but the three substantive clauses were carried by vote at the meeting on June 28. This is the bare record of the result of the proceedings, but in the course of the discussion Colonel Strange had defined the physics of Astronomy as including photographic, spectroscopic, actinic, photometric, and polariscopic observations of the sun ; ocular, photographic, and photometric observations of the moon, and observation and examination of planets, nebulæ, comets, zodiacal light, stars, asteroids, by the same methods so far as they are applicable. Further, he suggested the resolution that, " having in view the extent of the work above indicated, and the fact that no individual has as yet distinguished himself equally in these researches and in the more exact department of Astronomy, it appears to be both an administrative convenience and an intellectual necessity that the two departments should be kept distinct."

There was evidently a feeling against the establishment of a Solar Physical Observatory * under independent control apart

* Though it does not appear on the records, it is known that Mr. Lockyer's name was associated by many with the proposed Solar Observatory.

from Greenwich, but the opponents of an establishment of this kind were prepared to support an extension of the existing National Observatories.

The spirit in which the discussion was conducted and the feeling that existed in the Council may be inferred from the fact that at the meeting in 1872 November, Mr. De la Rue, Colonel Strange, and Mr. Lockyer retired from the Council; the first because he felt " that the opinion of the majority had diverged considerably from his own on various occasions " ; Colonel Strange, because he thought that certain members of the Council were incompetent and that these exercised an undue influence ; and Mr. Lockyer, because Mr. Proctor made repeated attacks on him in " certain obscure prints," and for this reason he did not wish to sit at the Council table with him. The secession of these three Members did not restore harmony, but the factious spirit had full play in the selection of a recipient for the Gold Medal in the follow-ing February. At the Council meeting in November it was proposed by Professor Pritchard, who was strongly supported by Dr. Huggins, that Mr. Lockyer, M. Janssen, and Professor Respighi should have the medal jointly, in accordance with the Bye-law passed in 1871 June. Mr. E. B. Denison (afterwards Sir Edmund Beckett, Bart., and finally Lord Grimthorpe) proposed that Mr. Proctor should have the medal for his contributions to astronomical literature, especially his Charts of Stars and theories about their distribution, and his papers on the Transit of Venus. Other names were proposed as recipients, but the contest was mainly between the supporters of the three solar observers first named, and those of Mr. Proctor. Professor Pritchard withdrew the name of Respighi before the discussion, but at a later stage—when he saw that there was much opposition to Mr. Lockyer—he was not allowed to withdraw the name of that gentleman, otherwise M. Janssen would have got the medal. There was a decided opinion among certain influential members of the Council that Mr. Proctor's work, though very voluminous and painstaking, did not deserve this high recognition, and it was eventually decided that the Gold Medal should not be awarded in 1873. It seems possible that Proctor had been proposed merely to set up a candidate in opposition to Lockyer.

An early opportunity for further dissension arose in the election of the Council in 1873 February, and on this occasion Colonel Strange was the avowed aggressor. The Council, according to custom, prepared a list of names to be submitted to the Society for election. A few days before the Annual Meeting a circular was issued to all the Fellows of the Society by Colonel Strange. In this he called attention to the recent resignations, which were not

the result of any mutual understanding, but were " due to dissatisfaction with the present composition of the Council." He said that
its members should be " competent for the work in hand, and
desirous of advancing it with single-minded earnestness." He
mentioned no names, but thought there would be little difficulty
in finding out to whom he alluded, and gave a long list of names,
from which a good choice of possible new members might be
made. A few days later (February 10), Colonel Strange issued a
balloting list in opposition to that of the Council, while the
President (Cayley) and the Secretaries (Dunkin and Proctor),
apparently without knowing of this rival list, sent out a reply to
Colonel Strange's circular. In this they merely pointed out that
his list of recommendations included some who had already served
on the Council or who did not wish to serve, while business considerations forbade extensive changes at a time when the Council
had recently lost the services of two chief officers.

The proceedings at the Annual General Meeting on 1873
February 14, consequent on the issue of these circulars, are thus
reported in the *Astronomical Register* (p. 66) :—

> A desultory conversation then ensued (after the appointment
> of the Scrutineers of the ballot), in which Professor Pritchard, Sir
> G. B. Airy, and Mr. Bidder were prominent, arising from the fact
> that Col. Strange had issued a circular to the Fellows expressing
> dissatisfaction with the list of Officers proposed by the Council, and
> afterwards had circulated an opposition list. . . . A long and
> stormy discussion ensued in a most crowded meeting, the principal
> speakers being Mr. E. B. Denison, Q.C., Mr. Chambers, Mr. Ranyard,
> Mr. Proctor, Col. Strange, Sir G. B. Airy, Dr. De la Rue, the Rev.
> T. Wiltshire, Mr. Balfour Stewart, and Mr. Bidder. Col. Strange
> was repeatedly challenged to substantiate his statements of the
> incompetence of the persons objected to, and their combination
> for party purposes, but contented himself by stating that it was
> merely his own personal opinion, and that he had every respect
> for the individuals in question. A resolution expressing regret
> for Col. Strange's action having been carried, the ballot took place
> and occupied nearly two hours before the result was ascertained.

The Officers and Council elected were those of the Council
list with one exception. The Report ends : " In consequence
of the anticipated length of meeting, the usual dinner did not
take place."

The above is a brief account of the proceedings at an Annual
General Meeting of a somewhat unusual character. That they
were wanting in dignity may be judged from a sentence in the
Astronomical Register for 1873 March. " The conduct of the
meeting was hardly creditable to the oldest scientific Society.

It was certainly a matter for regret that a learned and reverend Professor holding office in the Society should have allowed his feelings to get the better of his manners." Some recriminatory letters which were sent to the Fellows of the Royal Astronomical Society as sequel to these proceedings reveal the feeling that underlay their action. The first of these * was signed by John Browning, T. W. Burr, E. B. Denison, W. Noble, and R. A. Proctor, and in it they affirmed that Colonel Strange's circular had been issued with Mr. Lockyer's assistance, and implied that the attempt to eject them (the signatories) from the Council was a retort for the rejection of Colonel Strange's scheme for the establishment of a Solar Physical Observatory. To this, Colonel Strange replied by a letter,† in which he hinted that Mr. Proctor's name had been proposed as recipient of the medal in November by a clique of his personal friends on the Council, and for this reason he (Colonel Strange) was justified in proposing a reform of that body. Two further letters were sent to the Fellows, one by Mr. Proctor, the other by several other Fellows.‡

2. 1873–1874

The observation of the Transit of Venus across the sun's disc, which was to happen in 1874 December, was the subject of some discussion among astronomers in the year 1873, and as it led to an incident in the history of our Society which reflected somewhat upon its officers, it may be well to consider here the circumstances relating to the phenomenon which caused the discussion.

The event has already been alluded to as a matter of contention between Sir George Airy and Mr. Proctor, and several communications on the subject by the latter will be found in volume **33** of the *Monthly Notices*. To realise the point at issue it may be useful to remind the reader that there are two distinct methods of using the Transit of Venus for finding the sun's distance. In one, Halley's method, two observers in the northern and southern hemispheres respectively, note the whole time of duration of the passage of Venus across the disc of the sun, and from the difference of these times of duration, or in other words, from the difference of length of the chords of the sun's disc described by Venus as seen from the two stations the sun's distance is computed. The time of duration is affected by the motion of the observer due to the earth's rotation during the interval between ingress and egress of the planet, which may be several hours. The increase or decrease of the duration due to this cause, which makes the determination more accurate if it lengthens the difference of interval

* *Astron. Register*, **11**, 93.   † *Ibid.*, p. 95.   ‡ *Ibid.*, pp. 120 and 122.

due to north or south position of the observer, is a factor to be considered. Writing generally, the effect of the earth's rotation is to shorten the duration of a transit because a place on the hemisphere of the earth which is towards the sun moves lineally in the direction opposite to that in which Venus moves, and therefore hastens her motion across the disc, and for two observers in the same hemisphere, the shortening is less for the one who is nearer the pole, because his linear movement is less. But a case to be specially considered is that of an observer in the Arctic or Antarctic circle, according as the northern or southern pole of the earth is towards the sun, who, though on the sunlit hemisphere, is carried by the earth's rotation in the same direction as the planet. To such an observer the duration of transit is lengthened, and at a December transit, if the planet is south of the sun's centre, this lengthening of the passage for a southern observer, which is already greater than that for a northern, will be an advantage ; but if the geocentric transit is north, its duration as seen by the southern observer, which is less than that seen by the northern, will again be lengthened, and this will make the difference of the observed times less than it would otherwise have been, and hence it will be disadvantageous to observe the phenomenon from a station within the Antarctic circle, which will be beyond the pole, so to speak, at mid-transit.

A transit will happen in December, if Venus is at inferior conjunction and at the same time sufficiently near to the ascending node of her orbit. A transit will happen in June if similar circumstances occur, the planet being then near the descending node ; in general, transits happen in pairs, the individuals of a pair being at the same time of the year and separated by an interval of eight years.

At the first of a pair of December transits, Venus will have just passed the ascending node, and since the planet is then in north latitude the geocentric transit will be above the centre of the sun's disc. It will be found from consideration of the length of the synodic period of Venus, and her sidereal period, that the second transit of the pair at either node happens before her passage through the node, and the chord of geocentric transit will in this case be south of the centre of the disc. The southern observer in each case sees the planet displaced northward, but, as explained above, the advantage that can be gained by an observer in the Antarctic circle—the southern pole of the earth being in the sunlit hemisphere in December—cannot be maintained at the first of a pair of transits, since it lengthens the duration of the shorter transit, and hence shortens the difference. At the second of a pair of December transits, omitting the consequences of the earth's rotation, the southern observer would see a longer transit than the northern because of his position, and if he were in the Antarctic circle the rotation of the earth would increase the duration as seen by him, and the addition would be to the advantage of the determination. This has led to the view that Halley's method is not specially suitable for the first of a pair of December transits (and is similarly unsuitable for the first of a pair at the Descending Node in June). In his paper read before the Society in 1868 December " On the preparatory arrangements which will be necessary for efficient observation of the Transits of Venus in the years 1874 and 1882," Sir George Airy said " that for reasons he had given in earlier papers (1857 May and 1864 June) the method by observation of the interval in time between ingress and egress at each of two stations at least, on nearly opposite parts of the earth (on which method, exclusively, reliance was placed in the treatment of the observations of the transit of Venus in 1769) fails totally for the transit of 1874, and is embarrassed in 1882 with the difficulty of finding a proper station on the almost unknown southern continent." He therefore relied for the 1874 transit at least on the method known as Delisle's, the principle of that method being that whereas at some stations the ingress is seen accelerated, compared with geocentric ingress, at others, differing from the first in longitude, it is retarded, and similarly for the egress. The determination therefore depends on comparison of the time of either phase as seen from two stations, the difference of longitude of which must be precisely determined, the record being made in local time.

For the transit of 1874, Airy selected the neighbourhood of Honolulu, Rodriguez or Mauritius, Auckland Islands or New Zealand, and Alexandria ; Kerguelen's Island and Crozet's Island were mentioned as being favourable so far as astronomical circumstances were concerned, but objections were raised to these by the naval authorities who were consulted. In 1869 March, Mr. Proctor brought before the Society reasons to show that the method of duration of transit, or Halley's method, was not inapplicable to the transit of 1874, because as Venus was far north of the sun's centre and the chord therefore short, the shortening by parallax from a southern station would be so great that this would outweigh considerations of an adverse kind. He suggested Enderby Land (66° south latitude, longitude 50° E.) as a suitable station, but failing this that either Kerguelen's or Crozet's Island would be sufficiently suitable. In the same communication he also proposed alternative stations to those chosen by the Astronomer Royal for the observations by Delisle's method. Crozet's Island and Kerguelen's Land, with some other places, being considered suitable also for this method.

Mr. Proctor followed this up by another paper to be found in the *Monthly Notices* for June 1869,* but the subject dropped as a matter of controversy in the Society until the year 1873. A sum of £10,500 had been sanctioned by the Treasury in 1869 May, and voted by the House of Commons at the end of the Parliamentary session of that year to defray the expenses of providing instruments and making observations of the transit, and the Astronomer Royal proceeded with the preparation for the expeditions. A further sum of £5000 was added for photography, which was undertaken at the instigation of Mr. De la Rue, the Board of Visitors of the Royal Observatory, at their meeting of 1871 June 3, having passed a resolution that a grant of £5000 ought to be made to cover the cost of photographic apparatus and observations at all the stations. Sir George Airy expressed himself guardedly as to hope of success.

There was evidently a feeling in some quarters antagonistic to the official scheme. An article, apparently by Mr. Proctor, was published in the *Spectator* of 1873 February 8 on the subject, and on February 13, the day before the Annual General Meeting of our Society, when the stormy scene already mentioned happened, another appeared in *The Times*, of which Mr. Becket Denison was the avowed author, stating plainly the history of the matter and urging that sufficient attention had not been given to Mr. Proctor's papers and suggestions in planning the scheme of the observations. Mr. Proctor wrote a letter to *The Times* of February 20, in which he stated definitely the principal alterations that he proposed in Sir George Airy's plan. First, that one of the Antarctic stations it was proposed to utilise in 1882 should also be occupied in 1874, Possession Island, near Victoria Land, being the place indicated ; secondly, that the region of northern India for observing the " retarded egress " of Venus had been completely overlooked in the Astronomer Royal's researches ; thirdly, that the station selected for observing the " accelerated egress " was unfortunate in many respects, especially as Possession Island, which is suitable for observing the transit by Halley's method, is also very suitable for observing the phase referred to ; and lastly, that as Lord Lindsay was equipping an expedition to observe at Mauritius, it was not necessary for a Government expedition to go to Rodriguez, which was so near, but that the cost of this might be applied in providing for the Antarctic station. The Board of Admiralty sent these papers to Sir George Airy for his opinion, and at their Lordships' request his reply was communicated to our Society, and is printed in the *Monthly Notices* of 1873 March. The Astro-

---

* From the *Astronomical Register* it appears that this was read at the May meeting.

nomer Royal pointed out that whereas Halley's method may have had the advantage over Delisle's at a time when it was not possible to determine longitude with great accuracy, this was not the case at the present time, and after comparison of probable errors of different combinations and consideration of local circumstances, he declined to recommend that an expedition be sent to Enderby Land (on the Antarctic circle in about the same longitude as Kerguelen) on account of the severities the expedition would have to undergo, or to any station in the Antarctic Continent. He did not think it necessary to give up a selected station because a privately organised expedition would be near, and he was endeavouring to establish a photographic station in India. On 1873 March 22 a statement of the general plan was made in the House of Commons. Mr. Proctor's comments on the Astronomer Royal's reply are to be found in the same issue (March) of the *Monthly Notices*. He pointed out further, that although the two naval authorities had (in 1868) considered it quite possible to occupy Possession Island for the necessary length of time for the transit of 1882, now that he proposed it for the transit of 1874 the opinions they then expressed were lightly regarded. He also contributed two other long papers on the general subject to the same number, and another in April, with the significant title "Note on the Approaching Transit of Venus, with special reference to the probability of absolute failure through the want of a due number of southern stations." He asserted that at Kerguelen Island, which had been selected, bad weather was almost a certainty, and that the three other chosen southern stations were very inferior. Two short papers by him on the transit, which it is not necessary to describe, appeared in the May number.

At the meeting in June it was announced from the Chair that at the suggestion of the Board of Visitors, the Astronomer Royal had applied to Government for the means of organising parties of observers in the Southern Ocean, with the view of finding additional localities in the sub-Antarctic regions for observing the whole duration of the Transit of Venus. The precise effect of this seems to have been that the *Challenger* was to report on Heard or MacDonald's Island, about 300 miles south of Kerguelen's Island (longitude 70° E.), as a place of observation, whereas Mr. Proctor urged that there were several islands in the ocean south of New Zealand which would supply suitable stations, or that a landing might be made on the Antarctic Continent. A vigorous warfare of words was carried on in *The Times* on the subject between Admiral Richards, the Hydrographer, Mr. Denison, Mr. Proctor, and the author of an article in the *Edinburgh Review*. The line taken by the Hydrographer will be learned from the following extract from a letter by him (see *Ast. Reg.*, vol. **11**, p. 224) :—

> Enthusiasts no doubt there are who, however accomplished they may be as astronomers, are wanting and cannot but be deficient on many subjects which it is as necessary to take into account as astronomy in a question of this kind ; and hence we are told to send to the Antarctic Continent and to visit a variety of small rocky islets interspersed over the Southern Ocean at distances from each other varying from 1000 to nearly 4000 miles, many of which are actual myths, while on those which do exist it is certain that there is no anchorage for a ship, and that even landing would be generally impossible.

In the supplementary number of the *Monthly Notices,* Mr. Proctor published a chart of the Antarctic and sub-Antarctic regions which are suitable for observing the Transit of Venus in 1874, and in the accompanying letterpress wrote the following paragraph :—

> The chart requires no explanation beyond perhaps the remark that the islands in the less-known regions have been taken from ordinary atlases (after comparison of several), in preference to the Admiralty charts ; because, after certain withdrawals from opinions expressed in December 1868, one naturally feels doubtful about Admiralty statements which would appear to be variable according to official requirements. It did not seem well to insert any island, or group of islands, in the chart with some such note as " Here, if convenient to those in authority, there is an island," or " this group of islands can be regarded as a reality or a myth as may be required," and so on.

Mr. Proctor, it will be remembered, was Secretary of the Society at this time and, temporarily, Editor of its publications, because Professor Cayley, who was the actual Editor, was President of the Society. The supplementary number was therefore completely under his control,* as this, unlike others, does not receive the direct sanction of the Council, and it was naturally considered that he had made unworthy use of the opportunity thus presented to him.

Mr. Proctor left England in October to deliver a course of lectures in America, and at the November Meeting of the Council a letter was read from him resigning the Secretaryship of the Society. Mr. Ranyard was chosen to fill the office *pro tem.*, and

* A long paper by Mr. Proctor, " Statement of Views respecting the Sidereal Universe," illustrated by folding plates, and another by Mr. Waters, similarly illustrated, appeared in the same number. He thought it necessary to say that none of these charts had been engraved at the expense of the Society.

Mr. Dunkin offered to undertake the duties of Editor until after the next Annual General Meeting. At this meeting also a letter was received from the Hydrographer, Admiral Richards, complaining of the paragraph in Mr. Proctor's paper in the supplementary number. The Council expressed their regret in a letter to Admiral Richards, that the Editor should have so misused his office, and in the next issue of the *Monthly Notices* (1873 November) the following note was inserted by special order of the Council :—

> The attention of the Council has been directed to certain remarks made by their late Editor in paragraph 2, page 533 of the supplementary number, volume xxxiii. of the *Monthly Notices.* The Council have entered on their Minutes a resolution expressing their strong disapprobation of the paragraph referred to.

Mr. Proctor sent a letter from New York in December, expressing regret for the circumstances which led to the appearance of his paper on the transit in the supplementary number, and saying that if he had known earlier of the arrangements described by Sir George Airy at the November meeting,* the paper would not have appeared, and that he had not definitely proposed the occupation of any specified southern stations, but merely a search for such stations in due time. This was read as a paper at the meeting of the Society in 1874 January. A letter from him in *The Times* of February 6, dated from New York, January 16, criticising the Council's note in the November number, spoke of his paragraph in the supplementary number as " a carelessly worded jest." Mr. Proctor attended the meeting on 1874 May 8, and read an explanation, Sir George Airy, as Vice-President, being in the Chair. †

Mr. Proctor contributed occasional papers in later years, and was sometimes present at the meetings, but did not take an active part in the affairs of the Society, being engaged in writing and lecturing at home and abroad.

For the actual observation of the transit, the Astronomer Royal collected a body of observers from the naval and military services, with some civilians, who received instruction at the Royal Observatory in the practical details of observation and photography, several of whom became Fellows of the Society. The observing parties started to take up their stations in the early part of the

---

* At this meeting the Astronomer Royal gave an account of the preparations for observation of the Transit in which he said that a photographic station was to be established in the north of India, and that he had carefully considered the propriety of establishing an additional station on Kerguelen Island or on the Macdonald or Heard Islands, on the suitability of which the *Challenger* was to report.

† Mr. Proctor's explanation or apology was not received as a paper, but was printed as a fly-leaf, with a footnote, " To be added to Vol. xxxiii."

summer of 1874, that for Egypt being naturally somewhat later. When the observers returned after the transit they were all placed under the superintendence of Captain G. L. Tupman, R.M.A., who had been leader of the expedition to the Sandwich Islands, to complete their share of the reductions or the measurement of photographs. The examination and final reduction was entrusted to Captain Tupman.

3. 1874–1875

At the February meeting in 1874, Professor Adams was elected President in place of Professor Cayley, and Mr. Cowper Ranyard to the office of Secretary that he had held *pro tem.*, Mr. Dunkin being his colleague. The Gold Medal of the Society was awarded to Professor Simon Newcomb, for his researches on the orbits of Neptune and Uranus.

In volume **35** of the *Monthly Notices* (1874 November to 1875 June) there are many papers, prospective and retrospective, relating to the Transit of Venus. At the November meeting the Astronomer Royal gave a long account of the Reports he had received from the British expeditions as to their journeys and establishment at their stations, speaking specially of the observations for the determinations of longitude. He remarked that the general arrangement of stations was precisely as it had been from the beginning, although in some districts there had been expansions of the original plan tending to multiply the places of observations, these now being Egypt, the Sandwich Islands, the Island of Rodriguez, New Zealand, Kerguelen, the last of which was an addition to the original scheme. Information that the observing parties had arrived at these places had been received from all except the last-named, and at the meeting in December, Sir George Airy was able to announce that successful observations had been made from Egypt and India, and by various expeditions sent by foreign countries, but information had not been received from the other British stations.

It seemed fitting that the year of the Transit of Venus should be marked by the erection of a memorial to Jeremiah Horrocks, of Hoole, in Lancashire, who had predicted and observed the Transit of Venus of the year 1639. A petition was presented to Dean Stanley and the Chapter of Westminster, signed by the Astronomer Royal, the President, and several prominent Fellows of the Royal Astronomical Society, requesting permission to place in Westminster Abbey a tablet or some other memorial of Jeremiah Horrocks. The subject was brought to the notice of the Fellows of the Society at the June meeting, and a request for subscriptions, which it was thought well to limit to sums not exceeding a guinea, was made and responded to. Formal consent having been given by the Chapter, on the advice of Dean Stanley, who interested himself greatly in the matter, a block of marble with a curved surface was added to the monument of Conduitt, the relation of Newton, which stands on the north side of the west door of the Abbey, and on this an inscription was cut recording Horrocks' achievements. The memorial stone actually forms part of the Conduitt monument. The whole sum raised by subscription was £89, 16s., of which £51, 1s. was expended on the memorial, including a fee of £25 paid to the Chapter. The balance was handed to the Council of the Society, and, with an added sum, was invested, the interest to be devoted to the purchase of books for the Library; and this money now figures in the annual accounts as the Horrocks Fund, together with the interest on a sum of £500 bequeathed by the Rev. Charles Turnor in 1853 for a similar purpose.

Apart from the Transit of Venus, Coggia's Comet was the astronomical event of this year. This comet was discovered by M. Coggia, at Marseilles, on April 17, and became a splendid naked-eye object in the northern sky in June and July. It was observed very completely with the spectroscope by Huggins, Lockyer, and others, while drawings of the comet were made by various observers and published in the *Monthly Notices*, **35, 36.**

On 1874 November 13 the Society met for the first time in the rooms it now occupies. In 1868 and the years following, Burlington House was altered and enlarged by the addition of the east and west wings, to provide suitable quarters for those learned Societies who still occupied rooms in Somerset House, which were required for other purposes. Mr. Charles Barry was the Architect, and the interiors were arranged to suit the requirements of the Societies. The western wing, the southern portion of which was allotted to our Society, was completed in the first half of 1874, and the removal of the Library and other property of the Society was effected, principally under the direction of Mr. Dunkin, between the June and November meetings.

Between the November and December meetings the Society lost by death the valuable service of their Assistant Secretary, Mr. John Williams. Mr. Williams had been a Member of the Mathematical Society, and became a Fellow of the Royal Astronomical Society when the two were incorporated. On the death of Mr. Harris, Assistant Secretary, in 1846 April, Mr. Williams was chosen to succeed him, and therefore ceased to be a Fellow. He was a student of science of many kinds, but the chief work of his early life was the study of Egyptian hieroglyphics. From this he turned to Chinese, and attained great fame for his knowledge of that language. Several papers on the astronomy of the

Chinese will be found in the *Monthly Notices*, and in 1871 he published a book on *Observations of Comets*, from B.C. 611 to A.D. 1640, extracted from the Chinese annals, which was in no way a publication of the Society, but was privately printed by the author. Mr. Williams could not be persuaded to resign his post, even when he might well have done so, on the Society's removal to Burlington House, but the sudden death of his wife on November 10, after a union of fifty-two years, brought on symptoms of heart disease, and he died after much suffering, 1874 December 3, in his seventy-eighth year.

The business of the Society was carried on for a time by Mr. Edwin Dunkin, junior, son of the Secretary, and advertisements were inserted in several papers which brought twenty-eight applications. This number was reduced to five, and finally Mr. William Henry Wesley was appointed, and entered on his duties soon after the Annual Meeting. Referring for further particulars to the Obituary Notice in *M.N.*, **83**, we shall only mention here that Mr. Wesley was born in Derbyshire, 1841 August 23, and died in his rooms at Burlington House, 1922 October 17. He came to London in 1855 and was apprenticed to an engraver. From 1862 he did a great many drawings for Huxley, Herbert Spencer, Richard Owen, and other scientific men. He was persuaded by Mr. Ranyard to apply for the post of our Assistant Secretary, and when he had entered on his duties he threw himself into his work with great devotion and became a valuable official, to whom both the Society generally and individual Fellows owe a debt which can never be forgotten. His great artistic skill rendered considerable service to astronomy by his beautiful drawings of the corona from photographs of many eclipses and his carefully engraved charts of the Milky Way. It is a satisfaction to his many friends to remember that Wesley had the pleasure to view a total eclipse of the sun at the Algiers Observatory on 1900 May 28.

Reference to the Annual Accounts in the February numbers of the *Monthly Notices* shows that Mr. Wesley, in addition to comfortable lodgings, received a salary on appointment of £150 per year, which was increased to £225 by resolution of Council on 1879 November 14, and later to £250. From 1860 to 1881 a sum of £60 yearly was paid to Professor Cayley, who succeeded Professor Grant as Editor of the Society's publications.

At the Annual General Meeting in 1875 February the Gold Medal was awarded to Professor D'Arrest, Director of the Observatory of Copenhagen, for his observations of nebulæ. The reading of the Report at this meeting disclosed an incident which provoked some comment, for it appeared that instruments had been lent to the Eclipse Committee of the British Association for observation of the eclipse of 1871, and that the observing party in Australia had presented some instruments to Mr. Ellery, Director of the Melbourne Observatory, among them being three modern instruments of small value belonging to the Society. But the matter had already been dealt with by the Council, and the British Association naturally made restitution by procuring similar instruments and handing these to the Society.

The activity of the Society during the first five years of the decade may be judged from the size of the volumes of the *Monthly Notices*. Volume **30,** 1869 November to 1870 June, contains 226 pages; volume **35,** 1874 November to 1875 June, 410, whilst two of the intermediate volumes, **33** and **34,** have respectively 582 and 492. The extent and number of the notes on the progress of Astronomy in the February number of the *Monthly Notices* may perhaps be an index of the taste and energy of the Secretaries rather than an actual record of progress; but it is to be remarked that these notes in 1870 occupied only 15 pages, whilst in 1875 they filled 38, excluding the Reports of the Transit of Venus, but including a long Report on Meteoric Astronomy. The state of amateur Astronomy generally is to be judged from the reports from the private observatories that are to be found in the February numbers. Some of these were falling out of the ranks of active workers. Mr. Lassell had re-erected his 2-foot reflector at Maidenhead after his return from Malta, but did little with it after 1870. Mr. De la Rue's Observatory at Cranford, Middlesex, ceased to exist in 1873, when it was dismantled, and the large reflecting telescope and other instruments presented to the University of Oxford. Mr. Carrington had no active connection with the Society after 1865, but had built a new Observatory at Churt, near Farnham, in Surrey, of a somewhat unusual kind, described in volume **30** of the *Monthly Notices,* which he did not apply to any particular purpose. The places of these older astronomers were filled by several new workers in the science. Mr. Huggins's Observatory at Tulse Hill was growing in value and importance. Mr. Lockyer had established an Observatory, with spectroscopic laboratory attached, in the neighbourhood of St. John's Wood, in north-west London, the principal instruments being a 6-inch refractor by Cooke, and a seven prism spectroscope, the latter being replaced in 1873 by a speculum metal diffraction-grating by Rutherfurd, and with these he was observing the chromosphere and prominences and sun-spots, and doing necessary comparison laboratory work on the spectra of metals. He communicated the results to the Royal Society, but a full report of the work done at his Observatory is to be found in the

Monthly Notices for 1874 February.   Lord Lindsay's well-equipped Observatory at Dun Echt, of which Mr. David Gill was in charge till 1876, was practically completed in 1873, and description of the instruments, and of the work done with them, will be found in the February number of the *Monthly Notices* of 1873 and subsequent years.   The equipment of the Temple Observatory, founded at Rugby School about 1872 in memory of a former headmaster, consisted of an 8¼-inch equatorial by Alvan Clark, and a 12-inch reflector by With, which were used largely for educational purposes, but also by Mr. G. M. Seabroke for solar spectroscopic work, and by Mr. J. M. Wilson for observations of double stars.   An Observatory was established by Mr. Edward Crossley at Bermerside, Halifax, Yorks, where many observations of double stars were made in the course of years.

Colonel Tomline set up an Observatory at Orwell Park, Ipswich, in 1874, the principal instrument being an equatorially mounted refractor of 10 inches aperture, which was devoted to cometary work.   Mr. George Knott transferred his home and Observatory, in 1873, from Woodcroft, Cuckfield, Sussex, where he had observed double stars and variable stars since 1859, to another place in the neighbourhood, but this cannot be called a new Observatory. The Observatory at Birr Castle, Parsonstown, was in active work, Lord Rosse (fourth Earl) being engaged on an investigation with the 3-foot reflector of the heat radiated by the moon, whilst the 6-foot reflector was used for examining nebulæ and for making drawings of Jupiter.   Lord Rosse's researches on lunar heat were communicated to the Royal Society and published in *Philosophical Transactions*, 1873.   At Mr. Barclay's private Observatory at Leyton, Essex, established by him in 1854, it was chiefly double stars which were observed.   The Observatory of Mr. George Bishop, junior, that had been removed from Regent's Park to Twickenham on the death of Mr. Bishop, senior, in 1861, was closed in 1876.   Mr. R. S. Newall's 25-inch telescope, now at Cambridge Observatory, was completed and set up at Gateshead about 1871.   A telescope with object-glass of 21¼ inches aperture made by Mr. J. Buckingham, which had been exhibited in the Great Exhibition of 1862, was transferred by him from Walworth to a well-equipped Observatory at East Dulwich, London, S.E., and used for occasional planetary and other observations.*

Mr. E. J. Cooper, who had established a fully equipped Observatory at Markree Castle, County Sligo, in 1831, died in 1863,† and was succeeded in the estate by his nephew, Colonel E. H. Cooper.   The nephew had not the same active interest in Astro-

---

* This telescope is now at the City Observatory, Calton Hill, Edinburgh.
† The date is incorrectly given as 1872 in *Monthly Notices*, **63,** 197.

nomy as his uncle, but allowed the telescopes to lie idle until 1874, when he secured the services of Dr. W. Doberck. The Observatory of Mr. Charles Leeson Prince, at Crowborough, where he had an historic telescope with object-glass made by Tulley of 6·8 inches aperture, should also be mentioned.

A Government expedition was arranged by the Eclipse Committee of the Royal Society to observe the total solar eclipse of 1875 April 6, a sum of £1000 being granted for the purpose and aid given in other ways. The duration of totality was considerable in the Indian Ocean and in the neighbourhood of Siam. Stations in one of the Nicobar Islands and in Siam were occupied, but the weather was very unfavourable. According to the programme laid down by the Eclipse Committee an attempt was to have been made during totality at the Nicobar Station to register photographically the spectra given by the different layers of the chromosphere and coronal atmosphere by the aid of spectroscopes and prismatic cameras used in conjunction with telescopes. Clouds entirely prevented any results of this kind being obtained. Results were obtained only at Siam (by a party led by Dr. Schuster), where several ordinary photographs of the corona were secured with different times of exposure.

4. 1876–1878

At the Annual General Meeting of February 1876, Dr. William Huggins was elected President, and Lord Lindsay took his place as Foreign Secretary. Mr. E. B. Knobel, who has sat at the Council Table almost continuously until the present date, was for the first time on the Council. The Report disclosed the fact that the total number of Associates and Fellows of the Society at the end of the previous year exceeded 600 for the first time, 566 of them being Fellows, and of these seven had been members of the old Mathematical Society. In connection with the instruments a curious and somewhat unsatisfactory incident was reported. A 2¾-inch telescope of the Sheepshanks collection had been lent to the Rev. Jonathan Cape, and after his death his goods were sold by auction, and two astronomical telescopes were included in the sale, one of these being the property of the Society. This, however, was not sold, but was lost in some way, and the Society were not able to recover it. The incident led to an inquiry, and it appeared that there were other instruments formerly in the possession of the Society that could not be satisfactorily accounted for. New regulations were made in consequence, that loans should be for one year only, unless renewal is granted on fresh application.

The Gold Medal of the Society was awarded in 1876 to M. Le

Verrier for his theories of the four great planets, Jupiter, Saturn, Uranus, and Neptune, and for his tables of Jupiter and Saturn founded thereon.   The address was, very appropriately, delivered by Professor J. C. Adams, the President, but M. Le Verrier, who was in temporary ill-health, was not present.

The elections of Officers and Council of the Society in the succeeding years were marked by the exhibition of considerable factious spirit.   It is unnecessary to give a detailed account of these disputes ; the following will suffice :—

In the list prepared by the Council of officers proposed for election in 1877 February, Captain Abney was proposed as Secretary, with Mr. J. W. L. Glaisher, in the places of Mr. Dunkin and Mr. Ranyard, the former of whom was retiring by his own desire.   Abney at this time held a paid post in connection with the Science and Art Department at South Kensington, as Inspector of Local Schools ; and South Kensington and all its connections were anathema to the party in the Council who were considered the opposition.   An alternative balloting list was issued, in which (without his knowledge) Mr. W. H. M. Christie and Mr. Ranyard were proposed as Secretaries.   On the ballot being taken it was found that Glaisher and Ranyard were the elected Secretaries, and that Abney was not chosen in any capacity.   The circumstances of his rejection were peculiar.   Some Fellows at the meeting voted for him as a member of Council and not as Secretary, and though he received 49 votes for the Secretaryship, and 16 as member of Council, he was considered not to be elected a member of the governing body, but had to give way to Captain Noble, for whom only 47 votes were cast.   The anomaly caused considerable comment, and some attempt was made to amend the Bye-laws to prevent its recurrence, by declaring that votes given for an office should count as votes for the Council.

A Special General Meeting was held after the ordinary meeting in 1877 June, at which this and some other alterations in the Bye-Laws were to be considered.   But as a clear month's notice had not been given, and the Fellows were therefore unable to propose amendments, it was decided after a somewhat stormy discussion to defer the matter till the next Annual Meeting.   When this came round the Council withdrew the proposed alteration, and no such amendment of the Bye-Laws has ever been made. But it has since then been assumed, and agreed by vote in 1878 February, that votes for an office should be taken as votes for a seat on the Council.

In the Annual Report of the Council in 1877 February it was announced that Mr. R. C. Carrington had bequeathed to the Society the sum of £2000 Consolidated Annuities.   Also that

£200 had been received from the trustees of the estate of Mr. T. C. Janson, formerly a Fellow of the Society, who died in 1863. This legacy, which became payable on the death of his widow, had, in conformity with the wishes of the testator, been added to the Lee Fund for the relief of widows and orphans of deceased Fellows.* It was also announced that the Society had received after the death of Miss Anne Sheepshanks, an Honorary Member, who died in 1876, a gift of 192 volumes which had formed a portion of the library of her brother, the Rev. Richard Sheepshanks. Her chief claim to the gratitude of astronomers is, however, founded on her pecuniary beneficence to science, for it was at her expense that the exhibition at Cambridge bearing the name of Sheepshanks was founded, and that the meridian circle now in use at the Cambridge Observatory was provided. On the death of her brother, as a record of her admiration and affection for him and his work, she transferred £10,000 Consols to the Master, Fellows, and Scholars of Trinity College, Cambridge, the interest of which was to be devoted to the advancement of astronomical and kindred sciences in the University. One-sixth of the sum supplies the income of the Sheepshanks Exhibition, the remaining five-sixths is reserved for purposes connected with the Cambridge Observatory, either in payment of stipends or in purchase of instruments. In 1860 a further sum of £2000 was placed by Miss Sheepshanks in the hands of Mr. Airy, for providing a new transit-circle for Cambridge Observatory, and the instrument made by the firm of Troughton & Simms was mounted in 1870.

During the year 1876 the Society came into possession of several series of original sun-spot records in various ways. First, the original drawings of spots and manuscript books of observations made by Carrington between the years 1853 and 1871 were presented to the Society by Lord Lindsay. They had been sold by public auction and were bid for on behalf of the Society, but were purchased by a bookseller, from whom Lord Lindsay bought them. Secondly, the Rev. Frederick Howlett presented to the Society five volumes of sun-spot drawings made by him between 1869 and 1876; and thirdly, the widow of. Professor Selwyn gave to the Society a series of paper prints from the solar negatives taken at Ely under the superintendence of the Professor in the years 1863–73. The gift of Carrington's MSS. led to the insertion in the *Monthly Notices* for 1876 March of a short list of the sun-spot manuscripts in the possession of the Society,

* A further sum of £500 came into the possession of the Society in 1877 under the will of Mr. C. Lambert, whose name appears in the list of deceased Fellows in 1878, but of whom no biography is given. He left a sum of money to be distributed among scientific societies, and of this the Royal Astronomical received the amount named.

which go back in date to the year 1819, and to the publication (in May) of an account of how the Schwabe drawings were acquired by the Society. In the year 1864, Messrs. De la Rue and Balfour Stewart wrote to Schwabe to request that he would allow his original manuscripts to be sent to England. Schwabe was loth to part with them, but consented to do so on the condition that they should be sent back to him " at any time that I should be desirous of looking into them during the short time of life still left to me. I do not think that I shall have an occasion to avail myself of the permission asked for ; but permit me kindly to believe that it is in my power to do so. After my death you may consider the whole of the observations as the property of the Royal Astronomical Society."

Thirty-nine volumes of Schwabe's diaries (1825–67) were therefore sent to the Kew Observatory, and were transferred to the library at Burlington House in 1880. Mention of the subject of sun-spots recalls the fact that the Rev. Temple Chevallier presented to the Society in 1851 his valuable series of observations of spots made at Durham Observatory, bound in two volumes, which are now in the library. The writer of the obituary notice of Canon Chevallier remarked that he was the first to institute in England the regular continuous observation of the solar spots, and that his methods were afterwards adopted by Mr. Carrington, who had been an Assistant at Durham under him. These statements were contradicted by Carrington in a note in the *Monthly Notices* (**34**, 250), in which he pointed out that Harriot (1610–13) was the first English sun-spot observer, and that his (Carrington's) methods were quite independent of any that had been hitherto used.

The planet Mars was an object of interest for more than one reason in the year 1877. It was in opposition on September 6, and nearest the earth on September 9, the distance being 0·377 in astronomical units, which was an unusually close approach, though not the closest possible, for the distance had been only 0·373 at the opposition in 1845, as it will be again in 1924, these being the two minimum distances during the two centuries 1800–2000. The opportunity was made use of by Mr. David Gill for the determination of the solar parallax, as we shall presently relate, and at this apparition of the planet its two satellites were discovered by Asaph Hall.

Another incident of the year 1877 was the search for the supposed intra-mercurial planet (Vulcan) at the instigation of M. Le Verrier. It had happened not infrequently that a spot or some other marking had been seen on the sun, which the observer assumed might have been a planet whose orbit was within that of Mercury.

Though it was clear that in some cases the appearance was merely that of a sun-spot, five such observations made respectively in the years 1802, 1839, 1849, 1859, and 1862 were thought by Le Verrier to have represented passages of a single planet across the sun's disc, and he deduced from the observations the elements of the orbit of a planet which would, if it actually existed, be in transit on 1877 March 21, 22, or 23. Mr. Hind supported this by showing that a sixth observation—that made by Stark on 1819 October 9, was also consistent with the same orbit. M. Le Verrier sent a letter to our Society inviting their co-operation in testing the predictions by observation, and the Astronomer Royal, in consequence, sent telegrams to observatories in India, Australia, New Zealand, and North and South America, with a view to keeping a continuous watch on the sun on these days, photographic if possible, and he made a statement on the matter at the meeting on March 9. The total failure of the observations both in the opposite hemisphere and in our own, renders it certain that no such object crossed the sun's disc at the predicted time. The Rev. Stephen Perry at Stonyhurst and Mr. Rand Capron at Guildford kept careful and continuous watch during the three days, the weather being very favourable at both places, but saw no trace of the object they were seeking.

The observation of Mars, for determination of the solar parallax, by Mr. David Gill, from the Island of Ascension, at its opposition in 1877, was an important event of this session, and though the story and the connection of the Society with the undertaking are briefly told by Gill in his autobiography in the *History of the Cape Observatory*, it may be not out of place to give here a sketch of the incident and others in the early career of this famous astronomer, who had recently begun his association with the Society.

In 1872, Lord Lindsay, the only son of the Earl of Crawford and Balcarres, established an Observatory at the family seat at Dun Echt, of which he put Gill in charge. For observation of the Transit of Venus in 1874, Lord Lindsay organised an expedition to the Island of Mauritius at his own expense, the greater part of the arrangements and work, which involved the determination of the longitude of Mauritius, being assigned to Gill. The observations of Venus in Transit were made by heliometer, and it was proposed to take advantage of the opportunity afforded by having the instrument in position at Mauritius, to make observations of the planet Juno, which was in opposition on 1874 November 5, for determination of the solar parallax. Owing to unforeseen delay in the journey the first observation of the planet was not made until November 12, but sufficient observations were obtained to secure a result which proved to be $8''\cdot77\pm0''\cdot041$. On his journey home, at the request of the Chief of the Military Staff of the Khedive at Cairo, and with the consent of Lord Lindsay, he measured a base line near the great pyramid in connection with the projected survey of Egypt, and on his return to Dun Echt he was occupied during the years 1875 and 1876 in collecting and reducing some of the astronomical results of the expedition.

But in 1876, Gill resigned his position at Dun Echt, and, probably induced by his experience in the observation of Juno, determined to make similar observations of Mars at this opposition in 1877, when the geometrical conditions were exceptionally favourable. Observations would be made most suitably from a station near the Equator. He applied to Lord Lindsay for the loan of the heliometer provided that he could otherwise obtain the financial means for the expedition, and this was readily granted. In the autumn of 1876 he asked the Government Grant Committee of the Royal Society for a sum of £500 to enable him to carry out his scheme from St. Helena or the Island of Ascension, undertaking to defray any costs which might exceed £500 at his own charge. The Committee did not feel justified in granting so large a sum for one object, especially as they had many applications for other purposes, but recommended that the amount should be provided independently from Government funds. Feeling some uncertainty as to the success of this proposed course, and to present undue delay, Gill applied to the Council of our Society, who received his request very sympathetically. At the ordinary meeting of the Society on March 9, the Astronomer Royal brought the matter to the notice of the Fellows, urging the necessity of the observations and suggesting that the sum might be provided by a subscribed fund, to which he himself was prepared to contribute £20.

Dr. De la Rue made the suggestion that the money bequeathed by Mr. Carrington might be used for the purpose ; but failing that, he generously offered to subscribe £100. Finally, the requisite sum of £500 was voted by the Council from the Society's funds on the understanding that £250 would be repaid to the Society on the joint security of Lord Lindsay, Dr. De la Rue, and Mr. Spottiswoode, if that sum were not obtained from some other source. In 1878, however, this sum of £250 was provided from the Government Grant Fund of the Royal Society, and the acceptance of the generous offer of these gentlemen and others was not necessary.*

* By his will Sir David Gill, who died 1914 January 14, left the sum of £250 to the Society " in grateful remembrance of a like sum paid out of the funds of the Society in aid of my expedition to Ascension."

The preparations for the expedition were attended by an incident almost tragic. At the meeting in May, Lord Lindsay exhibited this heliometer that he had put at the disposal of Mr. Gill for the observations of the coming opposition of Mars, and in preparation for the exhibition the instrument was set up in the meeting-room at Burlington House a few days previously, with the polar axis in the position relative to the horizon that it would have at Ascension, latitude 8° S. In making a final necessary alteration a holding screw was turned which proved to be shorter than was to be expected and had run out; so there was nothing to keep down the lower end of the polar axis, and this with all it carried crashed to the ground. The instrument alighted on the eye-end, which was driven through the floor and snapped off, and by this means the shock of the fall was broken and less damage was done than might have been expected. After the meeting the heliometer tube was sent to Messrs. Troughton & Simms, while Messrs. Cooke & Sons and Mr. John Browning repaired the mounting. Within ten days all was again in order, the instrument tested, packed and stowed on board ship. Short reports from Mr. Gill during his stay at Ascension will be found at various places in the *Monthly Notices*,* and the definite results in volume **46** of the *Memoirs*. Mrs. Gill's book, *Six Months in Ascension, An Unscientific Account of a Scientific Expedition*, tells the story from another point of view. Mr. and Mrs. Gill returned to England in 1878 January.

The solar parallax was a prominent feature of the session 1877–78. Sir George Airy opened the meeting in 1877 November by announcing the value of the constant 8″·760 derived from the British observations of the Transit of Venus and published in the Parliamentary Report. Mr. Stone sent from the Cape a re-discussion of the same observations, interpreting the meaning of the word "contact" according to his own view, which was different to that adopted in the official reduction, and derived the considerably different value 8″·884. This was received at the meeting in March and led to some discussion, Mr. Gill objecting to so large a value on the ground that it was not consistent with Struve's constant of aberration and Cornu's value of the velocity of light. A second paper by Mr. Stone, to be found in the *Monthly Notices* for April, in which he discussed the observations of contact made at the Cape, apparently because he felt they had not been given sufficient weight, in the official reduction, two of them having been rejected, supported his previous result and led him to affirm that the value of the solar parallax must lie between 8″·84 and 8″·92. At the meeting in June, Captain Tupman, under

* Vol. **37**, 310–326; **38**, 1–11, 17–21, 57–58, 89–90.

whose superintendence the reduction of the British observations had been made at Greenwich, presented a result of the discussion of the solar parallax from observations made by British Colonial observers beyond those given in the Parliamentary Report, which produced a result considerably larger, lying between $8''\cdot82$ and $8''\cdot88$; and in the supplementary number there is to be found the values derived from the photographs, which were absurdly small and quite inadmissible, alternative figures being $8''\cdot240$ and $8''\cdot082$.

A value of this fundamental constant of Astronomy derived from observations of Mars at the opposition in 1877 was communicated to the Society in December of that year by Mr. Maxwell Hall, an amateur astronomer in Jamaica, who had determined the parallax by means of the displacement of the planet in right ascension when far east and far west of the meridian, measured by transits of the planet and comparison stars over the wires of a 4-inch equatorial. The resulting parallax was $8''\cdot79$, with a probable error of $\pm0''\cdot060$. The parallax found by Mr. Gill from his observations of Mars at Ascension was $8''\cdot78\pm0''\cdot012$.

Opportunity had been taken by several persons of the close approach of Mars at the 1877 opposition to make observations of its surface features, and several communications of this kind will be found in the *Monthly Notices* of this period—Mr. N. E. Green, an artist and amateur astronomer, had made a journey to Madeira for the purpose. His drawings of the planet, made in 1877 August and September, were presented to the Society in November and are published in volume **44** of the *Memoirs*. The outer satellite of Mars was observed in England by various observers.

There was a transit of Mercury of 1878 May 6, which was observed by many persons, features specially looked for being a halo of light round the dark disc of the planet in transit, and a white spot said to be sometimes seen at its centre, which were discussed at some length at the meeting in May. It appears that these were considered important circumstances, for though no time observations were made at Greenwich owing to the state of the sky at ingress and egress, the planet in transit was examined at the Royal Observatory for the physical features above mentioned.* Other communications to the Society that may be specially noted are two papers on the Lunar Theory by Professor Adams, and three by Mr. Neison.

Mention may also be made of the visit of Professor Rutherfurd, of New York, in 1878 May, who presented some photographs of the sun taken in 1871, showing granulations or rice grains, for the purpose of correcting to some extent a statement made at an earlier meeting that certain photographs taken by M. Janssen were the

* *M.N.*, **38**, 397.

first to show such markings; and secondly, a magnificent photograph of the solar spectrum which reached across the meeting-room. The original spectrum had been found by an interference plate, or grating, with about 17,300 lines to an inch, the machine, with which the lines were ruled having been constructed by himself. Some description of his method of procedure will be found in the report of the meeting in volume 2 of *The Observatory* Magazine. As a point in the history of Astronomy, it may not be out of place to record here that the first number of that periodical appeared in 1877 April, by initiation of, and under the Editorship of Mr. W. H. M. Christie, who had the support of many leading astronomers.

The following are extracts from the prospectus issued on a post card early in April :—

The Observatory. A Monthly Review of Astronomy

No. 1 will be published on April 20. The successive numbers will be forwarded post free to Subscribers on the third Friday of each month.

Subscriptions for three months (including postage), three shillings (for the United Kingdom).

The Observatory will aim at presenting in a popular form a general review of the progress of Astronomy, and at promoting the activity of observers by affording early intelligence of recent advances in the Science.

Contributions have been promised by Captain Abney, Sir G. B. Airy, Col. A. Campbell, Dr. De la Rue, Mr. Dunkin, Mr. J. W. L. Glaisher, Professor A. S. Herschel, Mr. Hind, Mr. Knobel, Mr. Knott, Rev. E. Ledger, Rev. R. Main, Mr. Neison, Professor Pritchard, Professor J. Stuart, Mr. J. M. Wilson.

The Gold Medal was not awarded in 1877, though the claims of several persons as recipient were considered at the Council table. Mr. Lockyer was selected as the recipient at the meeting of the Council in December (1876), but the selection was not confirmed at the January meeting. It seems remarkable that Mr. Lockyer should never have received the medal, considering how many times his name had been proposed ; but since a majority of three-fourths of those present is requisite for confirmation, it is not difficult to see that three or four opponents are sufficient to effect their joint purpose. In 1878 the medal was given to Baron Dembowski for his observations of double stars, which branch of Astronomy was being actively followed at the time, as in the Report of the Council in 1877 February on the Progress of Astronomy, five of the paragraphs related to double stars.

The similar Report for 1878 has notices on four researches on the lunar theory, one being Professor Newcomb's investigation of the motion of the moon by examination of observations of early eclipses and occultations; others were by Dr. G. W. Hill and Professor Adams on the motion of the moon's perigee and the motion of the node respectively, and the fourth the numerical lunar theory on which Sir George Airy had been engaged for several years, and on the progress of which he occasionally made reports. Mr. Neison communicated four papers in the session 1877–78, one on " Hansen's Terms of Long Period " being specially instructive. Meteoric Astronomy is represented in the Annual Reports of 1876 and 1877 by long notices prepared by Professor Alexander Herschel, who also published in the *Monthly Notices* for 1878 May a long list of known accordances between cometary and observed meteor showers.

It is made evident in several ways that the study of meteors was making progress. Mr. Denning, who was elected a Fellow of the Society in 1877 June had published several extensive lists of radiant points from observations by himself and others, and in 1878 January he contributed one which contained his first suggestion of Stationary Radiant Points. The satellites of Mars were naturally mentioned in the 1878 Report, and the Council were fortunate in receiving a graphic account from Professor Asaph Hall of the circumstances attending the discovery, which, as he explained, was the result of an organised search. The nova in Cygnus discovered by Schmidt at Athens on 1876 November 24, received little attention in its early stages from English astronomers, as immediate announcement of the discovery was not made. The Society took no concerted action and there was no grant from Government funds or those of other Societies for observing the total solar eclipse of 1878 July 29, when the track of totality passed across the western states of America, though several Fellows of the Society made the expedition at their own expense. Professor Thorpe and Dr. Schuster were the guests of Professor Asaph Hall. Mr. Lockyer took up his position with Professor Watson at a place known as Separation, on the Union Pacific Railway. The eclipse was well observed in several respects, a sensational item being the reported discovery by Professor Watson and Mr. Lewis Swift, independently, of an object of magnitude $4\frac{1}{2}$, near the sun, which was neither a known star nor a planet, and was supposed to be the long-sought-for intra-Mercurial planet, Vulcan.* This, however, has not proved to be the case. The appearance of the corona was of the kind now known as the minimum type, and this distinction between types appears to have arisen on this occasion, for Professor Young remarked, " The

* See *Observatory*, **2**, 161, 193, 235.

eclipse of 1878 has demonstrated that the unknown cause, whatever it may be, which produces the periodical sun-spots at intervals of about eleven years, also affects the coronal atmosphere of the sun." *   The corona in this eclipse gave a continuous spectrum, without any of the bright lines which had been seen on former occasions.   The 1474K line was conspicuous by its absence, which was established not only by eye observations, but also by photographs taken by Dr. Draper, Professor Harkness, and Mr. Lockyer, a diffraction-grating having been placed in front of the camera. The writer of a note in *The Observatory*, from which the above information is taken, goes on : " We trust that the success of this method will be a lesson to certain Fellows of the Royal Astronomical Society, who with glib assurance asserted that it was physically impossible to photograph the spectrum of the corona, and who did their best to prevent the method from being even tried under Mr. Lockyer's auspices in the eclipse at Siam." †

In the year 1878 a legal transaction took place somewhat to the financial benefit of the Society, the story of which is precisely described in the *Monthly Notices*, **39**, 211 (February 1879) :—

> In the year 1836 Dr. John Lee, who was then Treasurer of the Society, executed a deed of gift, by which he conveyed to the Society the advowson of Hartwell, and in 1844 by another deed of gift he conveyed the advowson of Stone, in Buckinghamshire, to the Society.
>
> Shortly before the date of these deeds his estate had been resettled by Private Act of Parliament, by which the manors of Hartwell and Stone were entailed, leaving Dr. Lee with only a life interest in them.   The living of Stone fell vacant in the lifetime of Dr. Lee, and the Society presented Dr. Booth, who held the living until last April, when he died.   Soon after the death of Dr. Booth the Council received formal notice that the present Lord of the Manor of Hartwell, Mr. Edward Lee, intended to dispute the Society's title to both livings, on the ground that they were (in legal language) appendant to the manors and consequently included in the entail.
>
> The Council nevertheless presented the Rev. James Challis, M.A., of Trinity College, Cambridge, who, besides being the son of Professor Challis, a distinguished Fellow of the Society, had many other recommendations.   Mr. Lee took steps to dispute the presentation, and the Council employed a solicitor who had unusual opportunities of being acquainted with such matters.   The result of this investigation was that he found that the advowson of Stone had long been severed from the manor, and that the Society's title was good ; but that in the case of Hartwell the matter was at least very doubtful.   The litigation to establish such a claim, depending on historical inquiries, would be very costly ; and even

* *Observatory*, **2**, 235.   † *Ibid.*, 162.

if the Society succeeded it would have to bear a great part of the costs, which could not, according to the rules laid down by the Courts, be recovered from the other side.

It was found that Mr. Lee did not wish to present any friend of his own to the vacant living, and was willing to settle the dispute by purchasing for the sum of £700 the advowson of Stone, subject to the Society's right of presentation to the existing vacancy, and to a release of the Society's claim to Hartwell being included.

These terms were submitted as a resolution at the Annual General Meeting held in 1879 February. It was agreed to, and the Rev. James Challis was presented to the living by the Society.*

5. 1879–1880

The outstanding incident of the session 1878–9 was again of a contentious nature, and brought a new personality on to the scene, who put the Council in a rather unpleasant position. This was Mr. Herbert Sadler, who had joined the Society in 1876 November immediately after leaving Cambridge. He appears not to have taken up any definite profession, but engaged himself in miscellaneous scientific work. He was specially devoted to double-star Astronomy, and by the aid of a $3\frac{1}{4}$-inch telescope, the property of the Society, added notes and corrections to published catalogues.

At the Council meeting in 1878 December, when, according to custom, the "House" Balloting List was prepared, Mr. Sadler's name was included, though he had made no contribution to the Proceedings of the Society up to that date, and though he was little known except to one or two members of the Council. He was elected in due course to the governing body in 1879 February. In January he contributed a paper under the title, " Notes on the late Admiral Smyth's 'Cycle of Celestial Objects,' Volume the Second, commonly known as the 'Bedford Catalogue,' " which was the cause of much offence. The form of this work is well known. The second volume consists of a list of double and multiple stars, which are described individually in a chatty and historical manner, and gives the position, angle, and distance of the companion, from the results of his own observations. Mr. Sadler asserted that many of these supposed positions were much in error, and insinuated that they were not the result of measures, but that Admiral Smyth had simply copied or " followed " the observations of previous observers. An example will explain the precise nature

* It appears from the obituary notice of Mr. Challis (*M.N.*, **80**, 345) that the parishes of Hartwell and Stone were united in 1902.

of the charge. In the Bedford Catalogue the position, angle, and distance of the companion of γ Persei are given as $226°\cdot0$, $55''\cdot0$. Mr. Burnham found the position-angle to be $324°\cdot1$, and Mr. Sadler suggested that Smyth had "followed" Sir John Herschel's position-angle of $224°\cdot9$, which is known to be a misprint. Similarly, Smyth gives the position-angle of the companion of δ Cancri as $163°\cdot0$, whereas later observers find $124°$ or $114°$, and the charge was made that Smyth copied a fallacious measure of Sir John Herschel's, $160°$. Mr. Sadler's paper consisted of a large collection of such examples, preceded by an introduction of criticism somewhat offensively worded. Mr. Sadler quoted from a letter by Mr. Burnham : "No publication of original observations in this or any other language can be named which contains so many serious errors. . . . There is no theory which will account for the many serious discrepancies. The measures generally agree substantially with those which are given from prior observers, but this agreement is kept up just the same where the earlier measures were all wrong." And then Mr. Sadler goes on in his own words : "As far as I am aware there is one catalogue only, and that not an original one, which surpasses the Bedford Catalogue in inaccuracy, and that Catalogue is the ' Reference Catalogue of Multiple and Double Stars,'" forming volume **40** of the *Memoirs*. This Reference Catalogue was one edited for the Society by the Rev. R. Main and Professor Pritchard from a manuscript left by Sir John Herschel. Mr. Sadler pointed out no specific mistakes in this catalogue, but merely made the general charge as above. Admiral Smyth had received the Gold Medal of the Society in 1845 for the publication of his " Cycle," the presentation having been made by Airy ; and Mr. Sadler called attention to what he called the cautious language used by the Astronomer Royal when speaking of the micrometrical measures. Airy's attention was drawn to the apparently slanderous statements by Mr. C. G. Talmage, who received the following letter in reply :—

1879 *February* 18.

I remember the award of the Medal of the Royal Astronomical Society to Captain Smyth, and my address on that occasion. I had great confidence in Captain Smyth's observations, but in regard to the award of the Medal, I thought that unusual circumstances required caution. 1. The award to a printed book, in many respects a mere catalogue, was unusual. 2. The language of the book is not satisfactory to strangers. 3. Captain Smyth was a member of the Council, and I know well that none of the voters for the Medal—that is co-fellows of Captain Smyth—had ever looked at the book.

ADMIRAL W. H. SMYTH
(1788–1865)

To face p. 202.]

With this account, you will perceive why I thought it proper that the Society should be furnished with all the vouchers for the observations.*

Airy appeared not to be particularly concerned about the attack on the Bedford Catalogue, because this consisted of specific charges that could be answered, but he objected to the loose and unsupported criticism of the " Reference Catalogue," and felt that the Council should have been more careful in publishing the paper. The matter was naturally the subject of much discussion privately. It appears that Mr. Sadler's paper had been referred in the usual manner to two Members of the Council; but one of these referees did not see the paper, and the other gave it merely a cursory examination.

At the meeting in March, Mr. Talmage was allowed to read " Remarks on Mr. Sadler's paper in the January number," consisting mainly of a letter addressed to him by Professor C. Piazzi Smyth, son of Admiral Smyth, in which the Professor naturally protested against Mr. Sadler's strictures. This gave an opportunity for several persons to state their views. A number of Fellows spoke vigorously against Mr. Sadler, and imputed to him a feeling of malignity and the desire to throw discredit on an honourable man. Others took a different line and were inclined to make excuses for Mr. Sadler, on the ground that his remarks were nothing more than fair criticism, and that it was advisable to correct errors in a published work. Mr. Sadler was not prepared to withdraw anything. He said that he had not imputed dishonesty to Admiral Smyth, that his remarks were legitimate criticism, and that he had the right to point out errors. On being pressed by Mr. Common to withdraw any imputation of bad faith, he said that it was impossible to deny the imputations for which he had given grounds in his paper.

Mr. Sadler's assertion that he had not imputed dishonesty to Captain Smyth is scarcely consistent with the fact that he had in his paper spoken of cases where Smyth " has presumably copied the measures of others." There also was in the *English Mechanic* on the day of this meeting (March 13) a letter by him, in which he spoke of the Bedford Catalogue in these words : " A stupendous fraud, as one of the first of double-star observers has happily termed it." The obnoxious phrase had been applied to the work by Mr. Burnham in a communication to the *English Mechanic*, which Mr. Proctor at a later stage endeavoured to soften by pointing out that the American use of the word " fraud " is

* In presenting the medal, Airy had asked Captain Smyth that the MSS. of his observations should be presented to the Society for reference. They were soon after deposited in the library.

something different to the English. Mr. Burnham endorsed this explanation in a paper contributed to the *Monthly Notices* in 1880.

The President (Lord Lindsay), Sir George Airy, and other Members of Council felt that some action should be taken to express their disapproval of Mr. Sadler's paper, for the publication of which they held themselves responsible. The Astronomer Royal prepared draft resolutions which were submitted to the Council at a Special Meeting on Monday, 1879 April 7. Of the essential resolutions, four in number, three only were adopted, and these after some amendment. The fourth resolution, which was condemnatory of Mr. Sadler's criticism of the " Reference Catalogue," was dropped. Airy took this emendation of his plan so seriously that he immediately, at this meeting on April 7, resigned his office of Vice-President of the Society, which led to the publication of the following note in *Nature* of April 10.

> In the interests of British science we have refrained now for some time from referring to the evil days which have fallen upon one of the most reputable of our learned societies. The time, how-ever, has now come when silence is impossible. At the meeting of the Royal Astronomical Society's Council yesterday, the Astronomer Royal, in consequence of the recent action of the Council—an action inevitable when the present constitution of that body is considered—resigned his seat at the board. We cannot too much regret that this Society, the traditions of which are second to none in Europe, should have been utilised for some years past by an advertising clique who have everything to gain by their connection with a body of honourable students of science. The withdrawal of men long known for their astronomical work from the Council commenced some time since. It has now culminated in the resignation of the Astronomer Royal, and we are informed that other resignations are to follow ; indeed, a man of scientific repute risks somewhat in being found among the Councillors. Surely the Fellows of the Royal Astronomical Society of London are strong enough to remedy such a state of things as this.

The three amended resolutions were considered and adopted by the Council at their ordinary monthly meeting on April 9 —(Good Friday fell on April 11 in this year)—when it was resolved that they should be further discussed at the next meeting of the Council, which would be on May 9. On that occasion extensive verbal alterations were made, and the fourth of the original resolutions proposed by Airy was restored with some modification. Finally, the resolutions given below were passed by the Council and read to the Fellows present at the meeting of the Society in the evening of the same date.

Since the publication of the *Monthly Notices* of the Society for January 1879, the attention of the Council has been recalled to an article headed, "Notes on the late Admiral Smyth's *Cycle of Celestial Objects*, etc., by Herbert Sadler, Esq.," containing remarks on several of the star measures given in that Catalogue, and also containing a sentence reflecting on the " Reference Catalogue of Multiple and Double Stars," forming vol. xl. of the *Memoirs*.

The Council, feeling themselves responsible for the contents of the Society's publications, cannot but express their regret that they should have authorised the printing of this article in its present form.

While they desire to uphold to the utmost perfect freedom in the criticism of scientific works, they would at the same time enforce a general rule to exclude from the Society's publications any imputation upon the personal honour or good faith of the authors : and they are sorry to observe in Mr. Sadler's article some remarks which are capable of being, and to the knowledge of the Council have been, construed in a sense which infringes this rule.

The Council are, moreover, of opinion that Mr. Sadler was not justified in passing a sweeping condemnation on the Reference Catalogue, which is irrelevant to the rest of the article, and is entirely unsupported by the citation of the instances on which his judgment was founded.

Mr. G. F. Chambers had given formal notice that he wished to propose the following resolution at this meeting in May : " That accusations of wilful fraud in connection with the composition by the late Admiral Smyth of his *Cycle of Celestial Objects*, having been made by Mr. Herbert Sadler, a member of the Council, the meeting is of opinion that the Council are deserving of censure for not having taken steps to have these charges either proved or retracted, and calls for the immediate resignation of Mr. Sadler." But after hearing the resolutions of the Council, Mr. Chambers said that he did not wish to proceed with his motion. Nevertheless, discussion followed, and the meeting was not conducted without some heat. It was on this occasion that Mr. Proctor, in a temperate speech, endeavoured to palliate Mr. Burnham's use of the word " fraud." Mr. Sadler's resignation from the Council was accepted in June.

It may be added that Airy, despite his unexpected resignation on April 7, was re-elected a Vice-President in 1880 February.

The question of the dependence to be placed on the measures contained in the Bedford Catalogue was settled once for all by two papers in the June number of the *Monthly Notices* for 1880. Mr. Burnham had measured most of the distant companions observed by Smyth, but by no one before him. Smyth's results were found " either roughly approximate or grossly incorrect." This paper was immediately followed by Mr. Knobel's " Notes on a paper entitled 'An Examination of the Double Star Measures of the Bedford Catalogue, by S. W. Burnham.'" Mr. Knobel had examined Smyth's original observing books belonging to our Society, and also the Spherical Crystal Micrometer with which a great many of Smyth's measures of position angles were made, and which was now in the possession of the Astronomer Royal. The principal results found were :—

1. Of the 150 objects re-observed by Burnham, 82 per cent. of the distances had weight one assigned to them by Smyth, which according to his own testimony represented "nearly worthless-ness." But these rough estimates, instead of being given as, for instance, $2\frac{1}{2}'$ or $1\frac{1}{4}'$, were printed as $150''\cdot0$ or $75''\cdot0$, which was very misleading.

2. Position angles were measured by pointing the double image of A in the direction of B, and as the axis of the spherical crystal could be revolved freely round the circle in either direction, two readings might be obtained for every position angle, and this was a fruitful source of error and confusion, which fully accounts for Smyth's apparent dependence on the results of his predecessors.

Mr. Knobel had thus completely succeeded in vindicating Smyth's character.

The Gold Medal was again not presented in 1880 February. Huggins was chosen as the recipient at the Council meeting in 1879 December, but the result was not confirmed with the requisite majority at the January meeting. Several other names had been proposed at the November meeting. Airy did not attend the meeting in December, when the claims of the nominees were discussed and voted on, but he sent his reasoned opinion on each to the President, Lord Lindsay. He supported Huggins wholeheartedly, because though he was not the original suggestor of the idea he was practically the inventor of the process for determining the velocity of stars in the line of sight, and " has arrived, and has shown other persons how to arrive, at most striking and important cosmical results."

The action of the Council was therefore in accordance with Airy's views in selecting Huggins. But apparently some of the members thought that they were being dictated to and were asked to bow to authority, and they therefore combined to frustrate the decision taken in December, as stated. In consequence of this, feeling again ran high, and the election of Officers and Council in February was again a contested one. Circulars and rival balloting lists were issued, among others by Professor Pritchard, not without some success.

Though it involves a slight departure from chronological order it will be convenient to complete now the history of the question of the Endowment of Research in so far as it affected the actual proceedings being issued of our Society.   The Duke of Devonshire's Commission on Scientific Instruction and Advancement of Science * continued its labours for some years, and by its action Mr. Norman Lockyer was given a post in the Royal College of Science in 1875, and a Committee was appointed by the Lords of the Committee of Council on Education in 1879, whose purpose it was to further the study of Solar Physics.   This Committee was composed of G. G. Stokes, Balfour Stewart, Richard Strachey, Norman Lockyer, W. de W. Abney, and J. F. D. Donnelly.   On its establishment Parliament voted an annual grant of £500, £300 of which was for payments to members of the Committee for fees and travelling expenses, leaving £200 for scientific assistance, chemicals, etc. It was arranged that in addition to this grant such assistance as was possible should be given by the general staff of the Department of Science and Art and in the Laboratory of the Science School.   This provision, scanty as it was, did not however pass without comment, and there was correspondence in the public press protesting against the application of public funds for the purposes of pure research.   The ordained staple work of the Committee appears to have been to arrange for keeping a photographic record of the solar surface; but a paragraph in a preliminary Report † made in 1880, to the Committee of Council, to determine whether the Treasury should be applied to for an extension of the vote for another year, shows how this simple programme was extended, and is informing in several ways.

> The Committee have had thirteen formal meetings.   In addition to this several Members of the Committee have carried on special branches of the inquiry; and Mr. Lockyer, as arranged when the Committee was appointed, has been charged with the general conduct of the observational and experimental work at South Kensington.   The Committee consider that by his Laboratory work and comparison of the results with Solar phenomena he has brought together a great body of evidence tending, *prima facie*, to conclusions of the utmost importance.   The labour and difficulty of the research are, however, so great that much additional time and attention must continue to be bestowed on it before the questions thus raised can be considered as finally settled; and the Committee think it of much importance that the researches now being carried on should not be interrupted.

The remainder of this Preliminary Report deals mainly with

* See above, p. 174.   † See *Nature*, 1880 May 13.

details about the taking of solar photographs, but contains the following paragraph :—

> As has been already explained, the Committee were appointed as a temporary measure to prepare the way for something of a more permanent and systematic nature, and it is to aid them in this work that the Indian observations have been asked for. What shape the research may permanently take it is impossible at the present time to predict.

The light in which the appointment of this Committee and its support by the Government was regarded by certain Fellows of our Society is clear from the account of the proceedings at the Annual General Meeting in 1881 February, when a motion, the text of which is given below, was proposed by Sir Edmund Beckett. The requisite notice was signed by Lord Crawford, Sir Edmund Beckett, Mr. Ranyard, Mr. G. P. Bidder, Captain Noble, and Mr. Barrow. Sir E. Beckett, in making the proposition, distinctly stated that he was not the originator of the movement.

> That a Meeting be held at the Society's rooms of the Members thereof, and such other persons as like to attend, to consider the question of the Endowment of Research by the Government ; and that the Astronomer Royal be requested to take the Chair at that Meeting.

The discussion on this motion took precedence of the reading of the Council Report by special resolution, and turned almost entirely on the question whether it was or was not *ultra vires* for the Society to hold a meeting for a purpose not definitely astronomical. Mr. Christie and Mr. De la Rue were among those who opposed the motion in this sense. It is clear from the reported speeches that there was considerable animus underlying this whole matter. In course of seconding the motion the Earl of Crawford said, "I will not go into any question of past or future party fights. I would advocate payment for results as much as you like ; but I do not think it is desirable in the interests of science, not only in this country but everywhere else, that men should be placed in a position with a fixed salary and really answerable to nobody for what they have to do." This seems to represent the feeling of the opposition.

The motion was amended to read : "That a meeting be held at the Society's rooms of members thereof to consider the question of the Endowment of Research by the Government," and carried, with a further resolution that the Council should fix a day for such a meeting. In consequence thereof the Secretaries issued a notice on March 21, calling a Special General Meeting on 1881 April 1 to consider this question, when the resolutions which follow, of which they had received proper notice, would be moved by one of the signatories.

1. That, in the opinion of this Society, the granting of public money for scientific research in cases where it does not appear that results useful to the public will be obtained, or where the researches proposed are likely to be undertaken by private individuals or public bodies, does not tend to the real advancement of science.

2. That this Meeting considers it inexpedient that a Physical Observatory should be founded at the national expense.

3. That this Meeting is of opinion that the Government grant to the Committee on Solar Physics at South Kensington should be discontinued.

4. That, in the opinion of this Meeting, full accounts should be published of all money expended by the Government for scientific purposes, and that in all cases the nature of the work to be undertaken should be defined as clearly as possible.

Signed by Crawford and Balcarres, Edmund Beckett, George P. Bidder, G. F. Chambers, J. Kennedy Esdaile, William Noble, A. Cowper Ranyard.

These are the names on the Notice given to the Secretaries. In the report of the meeting in the *Observatory* magazine, Airy's name is added.

A different view of the matter was put forward by Mr. A. A. Common in a circular sent to all the Fellows of the Society, dated March 27. In the course of this he said :—

If we take a broad view of the matter, and consider it fairly, what do we find ? That we, as a Society, formed, as our Charter says, " for promoting a general spirit of inquiry in Astronomical subjects," and as our first bye-law says, " instituted for the encouragement and promotion of Astronomy," are actually asked to pass resolutions that will effectually stop any supplies of money from the Government for any research that may not result in something useful to the public, or for the founding of a physical observatory, which will most certainly never be founded except by National money, and to stop supplies that already exist on a moderate scale. Our first object ought to be the advancement of Astronomy—this cannot be done without money. To stop the supply of money is to stop the advancement of Astronomy. It cannot be said that private individuals can do all that is necessary : there is nothing ignoble in receiving money from Government for such purposes as the founding of an Observatory, or for doing the many things that ought to be done.

Let us think how we stand with regard to other nations ; pass them through your mind ; we are all behind, doing nothing, and in a state of stagnation, and while America, France, Austria, and others are founding Observatories and promoting the Science in

a large and liberal manner, we are asked to assist in this state of
stagnation by leaving all in the hands of a few private individuals
who cannot, if they would, undertake the work that requires doing,
and now, because certain Fellows of the Society have an idea that
the *real* advancement of Science can only take place in a certain
way, we must lag behind and cease to lead as we did.

There was a crowded room on the occasion of the meeting on
April 1, and an account of the proceedings, with seemingly verbatim
reports of the speeches, will be found in the *Observatory* magazine
for 1881 May. Sir Edmund Beckett opened the discussion with
a speech of some length, speaking to the main question and giving
reasons why research should not be endowed. Captain Noble
seconded the resolution, but his speech is not recorded. Professor
H. J. S. Smith moved an amendment equivalent to the " previous
question," that " under existing circumstances there is no sufficient
reason for the expression of any opinion by the Royal Astronomical
Society in its corporate capacity upon the question of the Endowment of Research by the Government." He urged among other
things that the first resolution might be read as a censure on the
grant made by the Government to the Royal Society, and explained
that his words " Under existing circumstances " were intended
to guard against the possible event of Government funds being
at the disposal of persons who might be corruptible and use them
improperly, which he did not suggest for one moment was then the
case. Mr. Fletcher Moulton expressed somewhat the same views,
namely, that the subject was not one for discussion by the Society,
and mentioned in his speech that he had found that in a communication to the *English Mechanic*, Captain Noble had been
addressed formally as the " Secretary of the Society for opposing
the Endowment of Research," which is illuminating as to the
feeling on the matter. A letter from Sir George Airy to Captain
Noble, in the same capacity, giving the Astronomer Royal's
considered views on Endowment, will be found in the *Observatory*
for 1881 March, and in the *Athenæum* for 1881 February 19. Mr.
Ranyard, Mr. George Forbes, Mr. Schuster, and several others
spoke, and a written communication from Sir George Airy was
read, in which he said, " My objection to the establishment of the
Committee on Solar Physics is intended to apply only so far as
it is a paid Committee. I do not object to the purpose for which
the Committee was appointed or to the payment of a Secretary,
or to expenses incidental to an office. The Committee, I believe,
have faithfully discharged their understood duties, and I shall
willingly co-operate with them to the best of my power." Finally,
Professor Smith's amendment was put to the meeting and carried
by a majority of fifty to ten. Mr. Moulton moved, and Lord

Sir George Biddell Airy.

Engraved by C. H. Jeens from a Photograph.

London. Published by Macmillan & C?

Crawford seconded, that the debate be adjourned *sine die*, which after opposition from Mr. Ranyard was carried with two dissentients, and so the subject of the Endowment of Research passed from the purview of the Royal Astronomical Society. It is worthy of remark that Mr. Lockyer attended the ordinary meeting (which he had not done for some time) on May 13 and delivered a discourse on observations of the sun with the spectroscope, dealing specially with the observations made at South Kensington in the previous two and a half years.

In spite of the factious feeling that existed in the Society, which may be inferred from this history of the decade, and sometimes showed itself by rather unseemly conduct at the meeting, a considerable amount of valuable work was done, to which the volumes of the *Memoirs* (**37,** part 2, to **44** inclusive) and *Monthly Notices* (**30** to **39**) bear witness. In the former series there are (in addition to the reference catalogue of double stars in volume **40** and the bulky Eclipse Volume) some valuable papers by Cayley and Glaisher, the " Chronology of Star Catalogues " by Knobel, and a remarkably great number of observations of double stars by various observers. In the *Notices* we find first many papers representing the immense amount of labour expended on the Transit of Venus. Other papers deal as usual with a variety of subjects from most branches of Astronomy. Only Astrophysics is poorly represented, owing to the almost total absence of contributions from Huggins and Lockyer.

The material progress of the Society during the decade is shown by comparison of the figures in the Annual Reports of 1870 and 1880 February. In the earlier year the total number of Fellows was 512, as against 592 in 1880, the Associates numbering respectively 45 and 43. Of the 592, five had been members of the old Mathematical Society and paid no subscription. There is still an item in this tabular statement of the personnel of the Society headed " Non-resident Fellows." * It is a little unexpected to learn from the 1870 Report that there were at that date eleven such Fellows who must have joined the Society nearly forty years before. In 1880 the number had become reduced to four, and the last of these early Fellows died in 1885. The funded property of the Society had increased from £8400 to £13,200 (face value), and the income on the same from £237 in the 1870 account to £361 in that of 1880. In the report of the later year it is mentioned that the Library contained about 8000 volumes, and that since the removal to Burlington House nearly 3000 of these had been bound in a substantial manner, while there remained about 1000 to be similarly treated.

* See above, p. 80.

CHAPTER VII

1880–1920. (By J. L. E. Dreyer)

The history of the Society during the last forty years cannot yet be written in detail. The events are too recent, and most of those who were prominent members of the Society during the greater part of this period are still living, so that comments on their acts in a book published by the Society would be inappropriate. It must also be acknowledged that however fruitful the labours of the Society at large and of many individual members have been since 1880, and however vast the strides made by Science have been, the course run by the Society has been very smooth. When once the storm raised by the movement for " Endowment of Research " had subsided, its history was quite free from exciting episodes or controversies, which would supply good material to a historical writer.

For these reasons it has been decided not to deal separately with each of the last four decades, but to treat these forty years as one period, and chiefly to review those doings of the Council on behalf of the Society of which little or no printed record is accessible to the Fellows. After that we shall sketch what may be called the internal history of the Society and its publications. Owing to the enormous development of Astronomy in recent years and the great increase in the number of publications, we are not able in the limited space of this record to review the history of the Society in its connection with the progress of our science in the same detail as was done for the first sixty years of the century.

I

At the beginning of the seventeenth century the invention of the telescope supplied astronomers with an instrument which not only made it possible to get some idea of the nature of the heavenly bodies, but also very greatly increased the accuracy of observations of their positions. It is not too much to say that the rise of celestial photography towards the end of the nineteenth century effected an equally great revolution in Astronomy, though in this case the change came somewhat more gradually.   This is not the place to describe the development of the use of photography either in delineating the features of sun, moon, nèbulæ, and their spectra, or in providing new methods of determining stellar positions, which until the advent of the gelatine-emulsion process about the middle of the seventies made very slow progress. The immense increase in sensitiveness of the photographic plate then obtainable almost at once changed astronomical photography from a curious toy into a most important adjunct to an observatory. Draper led the way by photographing the nebula in Orion in 1881, but his early death in the following year left the field open to Common, whose brilliant success was, in 1884, rewarded by the bestowal of the Society's Gold Medal.

In 1883, Common proposed to Gould, who had photographed about seventy star-clusters at Cordoba, a joint arrangement for photographing the whole heavens.*   Gould's work in South America was so near its close that he was unable to undertake anything new, and the immense labour of measuring the plates would in any case have tended to deter him.   But by that time the problem of charting the stars by photography had attracted attention in various quarters.   The great number of stars visible on the Cape photographs of the great comet of 1882 showed the possibilities of the new method, and so did the experiments made in 1884 by the brothers Henry at Paris.   They found in the course of their continuation of Chacornac's Ecliptic Atlas, that their task became impossible when they approached the Milky Way, so that they were compelled to try the use of photography. In 1885 June, Admiral Mouchez, Director of the Paris Observatory, sent to our Society a cliché of part of the Milky Way as well as an enlarged photograph of the same.   The instrument employed had a specially constructed object-glass of 34 cm. aperture, made by Paul and Prosper Henry, and three exposures of an hour each were made, producing three images of every star, $4''\cdot5$ apart, to guard against accidental spots on the plate being mistaken for stars.   In the covering letter † it was suggested that six or eight observatories ought to combine in order to produce in the course of eight or ten years a complete set of maps of the whole heavens

Quite independently of the work done at Paris, Gill had in the meantime started work at the Cape, to continue by photography Argelander's and Schönfeld's *Durchmusterung* to the South Pole. The first plates were taken on 1885 April 2, and the work was completed by the end of 1890.   Thanks to the devoted co-opera-

---

* *The Observatory*, **9**, 326.

† *Monthly Notices*, **46**, 1.   Compare Mouchez' account, *Comptes Rendus*, 1885 May 11.

tion of Kapteyn, who measured all the plates, the publication of three volumes of the *Annals* of the Cape Observatory was completed in 1900, containing approximate places of 454,875 stars south of −18° Declination. A rare example of an immense piece of work carried out in a comparatively short time by the unwearied patience and perseverance of two individuals working at a great distance from each other.

In the meantime the French Académie des Sciences had invited scientific bodies and observatories in all countries to send delegates to an international astrophotographic congress, to be held in Paris to discuss the preparation of a photographic chart of the heavens. The congress was opened at the Paris Observatory on 1887 April 16. Of the eight members from Great Britain, three (Common, Knobel, Tennant) represented our Society, and a full report of the proceedings was printed in the Annual Report of the Council in the following February.* The following are the principal resolutions of the congress :—

A chart to be made of all stars down to the fourteenth magnitude, the plates to be in duplicate.

A second series of photographs with shorter exposure, including stars to the eleventh magnitude, to be made concurrently in order to form a catalogue and to determine fundamental positions in the first series.

The photographic telescope to be essentially similar to that used by MM. Henry.

After receiving the report of the delegates the Council appointed a deputation to wait on the Prime Minister (Lord Salisbury) and urge the desirability of this country taking part in this important undertaking. The deputation was, however, not received, as it was believed that it would not " lead to any profitable result." Eventually Greenwich Observatory was provided with a photographic refractor and enabled to undertake the zone +90° to +65°, while De la Rue generously presented the Oxford University Observatory with a similar instrument, with which the zone +31° to +25° Declination has been observed. These two Observatories have long ago finished their share of the *Carte du Ciel*, but several zones undertaken elsewhere are not yet completed. As the Society did nothing further, no more need be said.

The rapid rise of the application of photography has caused a great demand for negatives for serious study, for lantern slides for lectures, and prints for more casual examination or wall-decoration. The Society appointed a permanent Photographic Committee as early as 1887 June ; it has been regularly renewed every year. To include in our publications any large number of reproductions

* *Monthly Notices,* **48,** 212.

of the photographs in the possession of the Society was obviously impossible for financial reasons, but early in 1893 arrangements were made with Messrs. Eyre & Spottiswoode to copy such photographs and to sell copies at a fair price. Two years later it was decided to try as an experiment to let the Society act as a centre for receiving photographs intended for reproduction and distributing the prints, without being interested financially in the undertaking. This arrangement has been in force ever since and has been successfully managed by the Assistant Secretary, both prints (platinotype or aristotype) and slides being obtainable. Lists of the photographs in stock (or additions to such) are published annually in the Council Report. The collection had in 1920 February reached a total of 298 numbers, which looks as if " a long-felt want " had been filled by this arrangement. At the same time the pages of the *Monthly Notices* bear witness to the readiness of the Council to illustrate papers by a liberal application of photography whenever desirable.

A good many of the photographs issued are of total eclipses, and in their production the Society has had an active share. Though a vast number of eclipse results had been published by the Society, it had not taken a great part in the organisation of the expeditions. In 1887 the Council were invited by Professor Bredichin to send two observers to his summer residence near Kineshma, on the Volga, to observe the eclipse on August 19. The hospitable offer was accepted by Copeland and Perry on the nomination of the Council, but unfortunately bad weather prevented any results being obtained. In 1888 March, when the annual re-appointment of Committees by the Council took place, it was decided to appoint an Eclipse Committee. The Council of the Royal Society were duly notified of this, and the hope was expressed that as most of the members of the new Committee were already members of the Committee of the Royal Society, the work of the two bodies would be much facilitated. In the following year preparations were made by the R.A.S. Committee for observing the eclipse of 1889 December 21–22, and this eclipse, so far as British expeditions were concerned, was altogether managed by our Society.* An expedition under Father Perry was sent to Isles du Salut, near Cayenne; though very ill he managed to carry out his programme successfully, but died only five days later. Another expedition to Cape Ledo, in South Africa, failed altogether owing to bad weather.

Early in 1892, in reply to a letter from the Eclipse Committee, the Council of the Royal Society very cordially approved of the suggestion to form a Joint Solar Eclipse Committee. This was

---

* For full details about the preparations, the facilities granted by the Foreign Office and the Admiralty, etc., see *M.N.*, **50**, 2.

done at once, and the Committee thus constituted (which included the members of the Solar Physics Committee) arranged for observing the eclipse of 1893 April 16 in West Africa and in Brazil.*   In 1894 January the Council of the Royal Society enquired what the R.A.S. intended to do with regard to the eclipse of 1896, and whether they would take joint action with the Royal Society as in 1893.   To the suggestion that the Royal Society might appoint additional members of the R.A.S. Committee the Council of the senior Society not unnaturally objected.   They proposed instead of this that a permanent Joint Committee should be set up, having executive powers, electing its own Chairman and officers, applying on its own authority to the Government Grant Committee of the Royal Society, and taking care of instruments purchased out of any grant thus received.   This was at once agreed to, and the permanent Joint Committee held its first meeting on 1894 May 2.   It has existed ever since, the members being appointed annually in equal numbers (at present eleven) from each Society, to the great benefit of solar research, as the preparations for and the observations of total eclipses have been most efficiently organised.   Arrangements were made in 1898 for the publication of results ; the preliminary reports were printed in the *Proceedings of the Royal Society*, and a sufficient number of copies were supplied to be issued as appendices to the *Monthly Notices* at the expense of the R.A.S.   Similarly with the final Reports in the *Philosophical Transactions*.   This arrangement came, however, to an end in 1906, a subject to which we shall return when describing the Society's publications.

While thus taking a leading part in organising eclipse expeditions from this country, the Council continued as in former years to watch the state of efficiency of the *Nautical Almanac*.   Hind had been Superintendent since 1853 and held this post till the end of 1891.   Towards the end of this long term of office he was perhaps somewhat unwilling to make any changes of importance ; but in 1890 June the Council took the initiative, probably instigated by a paper read by Tennant two months earlier.   In this paper various changes were advocated, while attention was called to the limited number of apparent places of stars, which was much smaller than that of the star lists of the *Berliner Jahrbuch* and the *Connaissance des Temps*.†   A Committee was appointed (including the Superintendent) to report to the Council as to whether the Society should approach the Admiralty on the subject.   The Committee handed in their report in 1891 June and it was approved by the Council.   Among the improvements suggested were : a considerable increase in the number of

* Report in *M.N.*, **53**, 472.   † *M.N.*, **50**, 349.

star places, means to enable a computer to include the short-period terms of nutation in the star corrections, ephemerides of satellites as in the American Ephemeris, the beginning of the year to be taken as the moment when the sun's mean longitude is 280°. The Admiralty were addressed with a view to steps being taken to carry these modifications into effect; they were adopted by the new Superintendent, Downing, beginning with the year 1896.

In 1897 November an alarm was raised at the meeting of the Council, that the Society had not been consulted about certain changes introduced in the *Nautical Almanac* from 1901, in consequence of a conference of directors of national ephemerides, held at Paris in 1896. A Committee was appointed (including Downing), which at the meeting in December presented lengthy Minutes, pointing out that the initiative of the changes in 1834 and 1891 had been taken by the Society. The new changes included the adoption of new values of astronomical constants and the use of Newcomb's planetary tables, as well as a catalogue of more than 400 stars. The Committee considered that these changes seriously concerned the Society and astronomers in general, as confusion might result in case the changes were not adopted everywhere. A copy of this resolution was sent to the Hydrographer, with an expression of the hope that the Society was not in danger of losing the privilege of being consulted by the Admiralty. To this a reply was at once sent, acknowledging that a mistake had been made, and assuring the Society that "there was no intention on the part of anyone at the Admiralty to act otherwise than in accordance with precedent." In a further letter dated December 23 the Admiralty asked the opinion of the Council, as there was a divergence of opinion between the Astronomer Royal on the one hand and H.M. Astronomer at the Cape and the Superintendent of the *Nautical Almanac* on the other, as to adopting the resolutions of the Paris Conference. My Lords were much impressed with the value of the system of international exchange of certain calculations, which, they understood, would be much facilitated by the adoption of the same data.*

The appointment of another Committee was the result of this communication. They advised, that since the changes had already been introduced in the *Nautical Almanac* for 1901, they might be provisionally continued in the volume for 1902, pending a fuller consideration of the matter. This report was adopted by the Council, recommending that all necessary data for reducing

* This system of dividing a considerable portion of the computing between the various national ephemerides was further elaborated at a conference in Paris in 1911 October. Cf. *M.N.*, **72**, 342.

to the Struve-Peters constants be given in an appendix up to the end of 1906.

An abstract of the *Nautical Almanac*, called Part I., "containing such portions as are essential for navigation," had been published at the request of the Shipmasters' Society, the earliest one being for the year 1896; price one shilling. This little almanac contained the monthly part of the big one unaltered, so as to save the expense of setting up the type afresh, though this involved giving sailors various things they did not want. Also the noon-ephemerides for Venus, Mars, Jupiter, and Saturn, the eclipse section, and a few other items. Though most useful to sailors, it was felt that this publication was capable of improvement. In 1910 October the Admiralty therefore addressed the Council of the Society, pointing out that the necessity of frequently checking the position of a ship by observations was generally recognised, and that it was thought that the labour of computing should be still further reduced, while matter of no use to the seaman should be got rid of. A Committee was appointed, including Mr. Cowell, who had not long before succeeded to the post of Superintendent on the retirement for age of Mr. Downing. This Committee reported in 1911 February. In their recommendations the general principle was followed of only giving the data to that degree of accuracy which can be made practical use of in the best observations at sea. This was taken as $0^s\cdot1$ and $0'\cdot1$, with variation in one hour to $0^s\cdot01$ and $0'\cdot01$ respectively, though some exceptions to this rule were made. The right ascension of the mean sun, the declination of the sun, and the equation of time to be given for every two hours, so that no interpolation would be necessary.

These alterations were carried out from 1914.* In this new " Part I." almost nothing is taken unaltered from the real *Nautical Almanac* except the data referring to eclipses.

At the Annual General Meeting in 1911 February, Gill, after delivering the Presidential Address on presenting the Gold Medal to Dr. P. H. Cowell for his contributions to the Lunar Theory and Gravitational Astronomy, delivered a second address on some points connected with the subject-matter of the first one.† After considering which kind of observations of the moon are particularly wanted for the improvement of the Lunar Theory, and which might be discontinued, he turned to the subject of the *Nautical Almanac* office. He pointed out, that whereas two out of the three great subdivisions of Astronomy, astrometry and astrophysics, are dealt with in the two national observatories administered under the Admiralty, the study of the third subdivision,

---

* For further particulars see *The Observatory*, **35**, 245.
† *M.N.*, **71**, 380–385.

astrodynamics, is not provided for in any national establishment in this country.   The functions of the *Nautical Almanac* office are limited to the more or less mechanical production of the *Nautical Almanac* ; but in the working out of the mathematical theories of sun, moon, and planets, or in the building up of tables on the basis of such theories, that institution takes no part whatever. And yet the ephemerides in the *Nautical Almanac* are computed from these tables.   As the *Nautical Almanac* office was now in charge of a very distinguished worker in astrodynamics, the medallist of that day, it was more than ever desirable that this country should not be left entirely behind in these matters, but that an additional assistant of high mathematical attainments and a couple of computers be added to the existing staff of the office.

Gill was very much in earnest and was not content with merely throwing out a suggestion.   In May of the same year (1911) he brought the matter before the Council of our Society.   A small Committee was appointed, from which a report was received in 1912 January.   This followed very much the same lines as Gill's address, referring to the brilliant work of Newcomb in the office of the American Ephemeris,* and to the Lunar Tables founded on Delaunay's theory, recently brought out by the Bureau des Longitudes.   The Committee repeated the proposal made by Gill as to the enlargement of the *Nautical Almanac* office.   But the matter got no further ; Gill died in 1914 January, and six months later came—the deluge.   It is very much to be hoped that this question may be reopened under favourable conjunctures.

While on the subject of gravitational Astronomy we may refer to a piece of work initiated by the Council and carried out under its general supervision.   In the Greenwich Observations for 1859, places of the moon from Hansen's tables were given for midnight of every day on which the moon had been observed at Greenwich, from 1847 to 1858.   The comparison of these places with those of the *Nautical Almanac* gave the excess of Hansen's places over those of Burckhardt, and it was assumed that this might be adopted without sensible error for the time of observation, so that the differences of Hansen's places and the observed places might be found thereby.   The procedure was, however, not strictly correct, as the change of the quantity H—B in the course of some hours was by no means always insensible.   It seemed, therefore, desirable to make the calculation of the differences H—B more

---

* In 1897 January, when Newcomb was about to retire from the office of Superintendent of the American Ephemeris, the Council passed a resolution to the effect that they had learned with great regret of the possibility of his work on planetary tables and the lunar theory being interrupted, and desired to put on record their sense of the great importance to Astronomy of the completion of this work.   A copy of this resolution was sent to Newcomb.

complete by computing places by Hansen's tables for mean mid-night of *every* day during the years 1847–61, after which time Hansen's tables were used in the *Nautical Almanac.* Knowing thus H—B for every midnight, interpolation would give that quantity for the actual time of every observation, and thus the determination of the apparent error of Hansen's tables could be completed. The expense of these computations was met by a grant of £320 from the Government Grant Committee of the Royal Society, and the results were published as an appendix to volume **50** of the *Monthly Notices* (1890).

There are, of course, various kinds of ephemerides which can never find a place in the *Nautical Almanac,* nor in any similar publication. For a number of years, Marth had occasionally published ephemerides of satellites or for physical observations of the moon and the major planets. In 1882 June he was engaged by the Council to prepare such ephemerides regularly every year for publication in the *Monthly Notices.* This arrangement ter-minated in 1891, when the Council expressed their regret that they did not see their way to continue it any longer, but hoped that Marth would communicate the data on which his ephemerides had been computed. This, however, he declared himself unable to do at the time, but announced his readiness to continue the ephemerides without remuneration; and he did in fact supply continuations of most of them till his death in 1897. From 1898 till 1906 similar ephemerides were published by Mr. Crommelin. At the Paris Conference in 1911 it was resolved that ephemerides relative to the physical observations of the sun, moon, and planets should be calculated by the American Ephemeris, and this has been carried out from 1916.

The computation of ephemerides is often the first step taken after the announcement of an astronomical discovery, and it is therefore natural in this place to allude to the arrangements made for the rapid distribution of news of discoveries. The first attempt at such in this country was made by the late Lord Crawford, from whose Observatory the " Dun Echt Circulars " were issued from 1879 to 1889. Similar circulars were sent out by our Society in 1880–82, about fifteen in all. When the " Centralstelle " for the despatch of astronomical telegrams was started at Kiel, it was suggested to let the Society act as an intermediate station for this country, but this was found to be impracticable and was never carried out. The central office at Kiel continued to have charge of this matter until the war put an end to its activity. In 1919 the International Astronomical Union established a central bureau for astronomical telegrams at the Brussels Observatory.

While the observatories in the United Kingdom, as a rule,

do not require to appeal for support to the Council of the Society, the case is different with those situated in the southern hemisphere. In some ways their responsibilities are greater ; they are very few in number, and the field of work open to them is an immense one, while some of them are dependent on local bodies for their very existence. This is, of course, not the case with the Cape Observatory, which as a Royal Observatory is only dependent on the Home Government, while Gill seems always to have been *persona grata* with the Admiralty, and to have understood the art of loosening the purse-strings. Yet even he found it on one occasion desirable to ask for the support of the Council. In 1892 he strongly urged the desirability of a Board of Visitors of the Cape Observatory being constituted, to meet once a year in London, and to consist of six members. He enlisted the sympathy of Lord Kelvin, at that time President of the Royal Society, who suggested that the Council of the R.A.S. should make a representation to the Admiralty on the subject. This was not the first time such a Board had been thought of. In consequence of the defective state of the Paramatta Observatory, Airy wrote to Sir Robert Peel in 1846 April, raising the question of a General Superintending Board for Colonial Observatories in order to maintain them in a creditable state. The Council now, in 1892 November, passed a resolution, embodying that passed by the Greenwich Board of Visitors in 1846 : that the Admiralty be urged to enlarge the powers of the Greenwich Board, extending them to the Cape Observatory by constituting a Committee (including the Astronomer Royal) to require reports from H.M. Astronomer, and making suggestions to the Government, in order to give greater unity to the work done at Greenwich and at the Cape.

This was duly sent to the Admiralty with the concurrence of Lord Kelvin. Nothing had come of the proposal in 1846, and nothing came of it now ; the Admiralty replying that they did not see the use of a Board which could never visit the Observatory. Gill was, however, empowered in future to refer to the Council of the R.A.S. for its support in connection with any important proposals. Annual reports were to be prepared by him in future.

The remarkably great accuracy of the heliometer observations of the minor planet Victoria in 1889 appeared to Gill to show that the time had arrived when astronomers should reconsider their methods of determining the positions of the planets. In the case of Victoria, the highest precision had only been required in the *relative* positions of the comparison stars to each other and of the planet to the stars. But for determining absolute positions of planets, we require in addition a fundamental system of star-places ultimately based on observations of the sun, and here the elimination of systematic errors is of vital importance. In a paper published in volume **54** of the *Monthly Notices* (p. 344, 1894 April), Gill discussed various points connected with this subject and asked whether steps should be taken for a more complete organisation of astronomical work, and whether it was desirable to hold an international congress to make preliminary arrangements. This paper was considered at two meetings of the Council in 1894, and various astronomers at home and abroad were consulted. In the end it was decided that it was not advisable for the Society to convene a congress on astronomy of precision.

In the endeavours made from time to time during recent years to put the Australian Observatories on a more satisfactory footing and to establish an Observatory for solar research in New Zealand, the Society has not been called upon for advice or support. The question has also been raised whether a Solar Observatory ought to be established in Australia, and the advocates of this movement took advantage of the presence in Australia of members of the British expedition to observe the eclipse of 1911 April 11 in the Tonga Islands, to call attention to the subject. Father Cortie, chief of the expedition, visited and reported very favourably on a proposed site.* A joint deputation of the British Association and the Royal Astronomical Society tried to take advantage of the presence in London of Mr. Fisher, Premier of the Commonwealth, but though he was unavoidably prevented from receiving them, they interviewed another member of his Cabinet and pointed out the importance for solar research of filling the gap in longitude, which could only be obviated by an Australian Solar Observatory. The excellent climatic conditions were also emphasised.†

Solar research in another favourable locality outside Europe has also been encouraged by the Society. In 1915, when Mr. Evershed undertook his first expedition to Kashmir to investigate the suitability of the climatic conditions for solar work, the Council was willing to co-operate by joining the Indian Government in sending out an independent observer of distinction. But as he was unable to be absent from his post for the necessary length of time, the proposal fell through.

The help of the Council was invoked in 1905 on the troublesome subject of lunar nomenclature. In his third paper on the determination of selenographical positions,‡ Saunder drew atten-

---

* *The Observatory*, **34**, 360.

† Since this account was sent to the printer, the welcome news has been received that a Solar Observatory has been sanctioned by the Government, and the appointment of a Director is imminently expected.

‡ *Memoirs*, **57**, 47. For a fuller statement of the difficulties see *Monthly Notices*, **60**, 41.

tion to the utter insufficiency of the system hitherto used. Lunar
formations were distinguished either by separate names or by
the names of adjacent formations with a letter added. But the
limits of the districts to which the particular names should apply
had never been clearly defined, so that it often happened that
when a letter is found on a map it was difficult to determine to
which of several names it should be attached. Again, many
selenographers considered themselves entitled to add fresh names
to the map, and as there was no recognised authority, it had happened that the same name has been given to two different formations, or that the same formation had received two different
names. It was therefore resolved by the Council (on the motion
of Professor Turner) that an International Committee on Lunar
Nomenclature was desirable, and that the Council would learn
with satisfaction that the International Association of Academies
would be prepared to appoint one. Saunder's representations
were also supported by the Royal Society and reached the International Association at its meeting at Vienna in 1907, when a
Committee on Lunar Nomenclature was appointed. Of the work
done by this Committee this is not the place to speak, but two
publications instigated by it (*Die Randlandschaften des Mondes*,
by Franz, and the *Collated List of Lunar Formations*, by Miss
Blagg) show how opportune the appointment of the Committee
had been. It was further determined to make an accurate map
of the moon as a vehicle for the authoritative names, the inner
portions of which were drawn by the skilful hands of Mr. Wesley,
our Assistant Secretary. The outer portions were to have been
drawn by Franz; since his death they have been drawn by
Miss Blagg, to Mr. Wesley's entire satisfaction.

Geodesy being closely connected with astronomical work, it
is natural that the Council should be warmly interested in the
geodetic operations carried on in this country. In 1904 May
attention was drawn by Major Hills to the alleged inferiority of
English geodetic measures to those of neighbouring parts of the
continent; and it was proposed that the Council should take
steps to obtain a reobservation of a geodetic arc in this country.
A resolution was passed a month later, stating that the Council
felt strongly the need of meridional and longitudinal arcs being
remeasured in the United Kingdom. The question was again
brought forward by Major Hills at the York meeting of the British
Association in 1906; and again at the Dublin meeting in 1908
there was a discussion on the proposed remeasurement. Soon
after, Major Close * suggested to the Council of the Association
that, before definitely accepting the view that the linear errors

* Director-General of the Ordnance Survey, 1911–1922.

of the British triangulation rendered a remeasurement desirable, it would be well to measure a base and remeasure a portion of the old triangulation remote from the principal bases at Salisbury Plain and Lough Foyle, in order to ascertain in a previously untested region what linear errors had accumulated. This was approved by the Council of the Association, and on its recommendation the Board of Agriculture and Fisheries authorised the Ordnance Survey to undertake the work. It was carried out in the years 1909 to 1912, and consisted in the measurement of a base on the southern shore of the Moray Firth, east of the town of Lossiemouth, and a test triangulation in Morayshire.* The investigation of these measurements showed that the linear errors of the old net-work triangulation of the United Kingdom are in the same terms as those to be expected in modern triangulation carried out in chains over similar distances. The error to be expected may be said to be of the order of one inch in one mile. The discussion showed that the excellent agreement between the measured and computed values of the bases at Lough Foyle and Salisbury Plain was not accidental, but is confirmed by the inter-comparisons of these bases with those measured at Paris and at Lossiemouth. These comparisons give results which are in some cases better and in some cases worse than those derived from modern work in Europe, India, South Africa, and the United States. The influence on modern figures of the earth of any remeasurement of the British arcs would be insignificant.

In 1907 a large surveying party was making a topographical survey of the area north of the German territory in Central Africa, near the Equator, along the 30th meridian. In April of that year Gill wrote to the Council that it was very important to retain this party for four months longer, in order to make a reconnaisance for the measurement of a geodetic arc of $2\frac{1}{2}°$ along the 30th meridian, forming part of the great African arc. The Berlin Academy had approached the German Government for funds to continue the arc east of Lake Tanganyika, which proposal (Gill suggested) would be much strengthened if it could be shown that work had been commenced north of the German territory. It was therefore proposed that the Council should give £50, to be added to £950 from the Royal Geographical Society, the British Association, and the Royal Society, after which the Government would be asked for another £1000. This was willingly agreed to by the Council. Eventually it turned out that owing to unfavourable weather the cost was nearly twice as great as estimated, and the Council in 1910 paid £25 towards covering the deficiency.

* Ordnance Survey, Professional Papers, new series, Nos. 1 and 2. London, 1912–13, 4°.

In this connection we may mention another case, where the Society threw its influence in the scales to further the promotion of an object associated with Geodesy. In 1903 March, Chandler addressed a letter to the Council, urging the desirability of establishing a southern belt of stations for the observation of variation of latitude. The Observatories at Sydney and the Cape being almost in the same latitude (differing only 4′) would furnish two stations for the investigation of Kimura's term, while a third one should be set up thirty miles south of Santiago.* The Council at once declared themselves much impressed with the importance of establishing a belt of southern latitude stations for two years. It appears, however, that the Central International Geodetic Bureau had already in 1896, six years before Mr. Kimura announced his new term, recommended that latitude observations be made on the parallel of Sydney. Eventually two stations were set up on the parallel − 31°55′, in West Australia and the Argentine.

Nearly at the end of the first century of its existence the Council was called upon to give an opinion on a subject of great importance, the commencement of the astronomical day. It was not the first time that this question had been laid before them. Since the days of Ptolemy astronomers had counted their day from noon, without the rest of the world being aware, that here was an intolerable grievance which ought to be redressed. But at the Washington Prime Meridian Conference in 1884 it was resolved among other things, " almost without debate, certainly without adequate consideration," says Newcomb, † that the astronomical day ought to begin at midnight like the civil one. This resolution, however, met with very little support outside the Conference. In a paper printed in the *Monthly Notices* in the following January, Newcomb expressed himself very strongly against the proposed change, chiefly on account of the discontinuity it would introduce into astronomical tables and ephemerides,‡ and at the Geneva meeting of the *Astronomische Gesellschaft* in 1885 August he did the same. § As he was Editor of the American Ephemeris, his vote was of great practical importance, apart from the weight any utterance of his would naturally carry. Of prominent Continental astronomers, Otto Struve and Oppolzer ‖ seem to have been the only ones in favour of the proposed change.

The matter came before the Council of our Society in 1885 June, when a letter from the Science and Art Department was read,

---

* *M.N.*, **63**, 294.   † *Reminiscences of an Astronomer*, p. 227.
‡ *M.N.*, **44**, 122.   § *Vierteljahrsschrift d. a. G.*, **20**, 228.
‖ *M.N.*, **44**, 295.

stating that the Committee of Council on Education had asked a Committee (including Adams, Christie, and Hind) to advise them. In accordance with the recommendations of this Committee a copy of the resolutions of the Washington Conference was now forwarded, "which My Lords consider to commend themselves for adoption." It was resolved on the motion of Christie that the Council of the R.A.S. concurred in those resolutions, and proposed that the change be adopted in the *Nautical Almanac* for 1890.*

But the opposition had been too strong, and it was probably realised that no other ephemeris would follow the example of the *Nautical Almanac*. The matter was therefore dropped and nothing more was heard of it for many years.

In 1917 attention was again drawn to the fact that the system hitherto in use was by many sailors found to be a fruitful source of error, and in 1918 January the Admiralty addressed a letter to the Society, requesting them to ascertain the views of astronomers about changing the commencement of the astronomical day. A Committee was appointed and reported in the following November. Of seventeen replies received to a circular, nine were decidedly in favour of the change, three decidedly against it, five not very decided. A specially important favourable reply had been received from the American Astronomical Society, who had consulted four observatories and eighteen individuals ; the result was three votes to one in favour of the change. The Committee recommended the change to be made in the *Nautical Almanac* from 1925, and their Report was adopted by the Council. The Admiralty have accepted this decision, and as the *Connaissance des Temps* and the *American Ephemeris* will make the change from the same date, astronomers are now committed to this innovation, for better for worse.

Among undertakings which have to be carried out by international co-operation, the cataloguing of scientific literature must of necessity be one, on account of the enormous mass of papers published annually. In 1895 December the Royal Society informed the R.A.S. Council that they had been requested by their International Catalogue Committee to arrange that each paper in the *Philosophical Transactions* and *Proceedings* should be accompanied by a statement of its contents, which would serve for use in the preparation of a subject-index. This plan was never adopted by our Society, but correspondence on the subject was carried on from time to time with the Royal Society. [At

---

* Airy apparently did not concur, as he wrote to Newcomb (no date given): " I hope you will succeed in having its adoption postponed until 1900, and when 1900 comes, I hope you will further succeed in having it again postponed until the year 2000." (Newcomb, *Reminiscences*, p. 227).

last, in 1900, the Council agreed to take part in an international catalogue ; but the schedule of classification, about which the Council had not been consulted, was modified by a Committee appointed for the purpose. It was also decided that the cataloguing of British astronomical literature should be done by one person with assistance from an Advisory Committee, and that he should be paid £30 a year from the Society's funds. This was accordingly done from 1901 to 1914 inclusive, when the arrangement was discontinued. Early in 1918 a letter was received from the Conjoint Board of Scientific Societies, enquiring whether the Society would be prepared to co-operate in forming the astronomical section of a proposed subject catalogue of scientific literature, as it was more practicable for this to be done by a single country than by an international organisation. But the Council decided that the value of a subject-index was not sufficiently great to justify the considerable expense it would involve, and that the Council was therefore not prepared to undertake the work.

We may mention here that the Conjoint Board was organised in 1916 for the purpose of promoting co-operation in appealing to the Government on matters relating to science, industry, and education. Our Society agreed to join as one of the constituent societies, and was represented on the Board by two members till 1922 April, when the Council decided to withdraw from the Board.

But the Society is much more intimately associated with another co-operative organisation established in 1917. A strong Committee was appointed by the British Association to arrange for meetings for discussion of the various branches of geophysics. It was felt that though this subject is closely connected with several important sciences, there is not any Society in this country for promoting its interests or bringing workers on geophysical subjects into contact with each other. The first meeting was held in the rooms of our Society on Wednesday, 1917 November 7, when the Chair was taken by the Astronomer Royal.* He pointed out that the sciences concerned covered a wide field ; progress in several depended to some extent upon public departments as well as upon the private investigator, and there was a danger that the latter might not be acquainted with what was being done by the former. It was hoped that these meetings would keep scientific interest alive in the work done on these subjects, and keep workers in touch with researches in geophysics in various parts of the British Empire, much of which was published only in official reports. A further service the meetings would perform

* See report in *The Observatory*, **40**, 444.

would be to enable workers in one branch of geophysics to obtain a general knowledge of the other branches.

These meetings have been continued five or six times during the session, the subjects discussed being magnetism, constitution of the atmosphere, aurora, geodesy, seismology, the earth's interior, etc. At the meeting of the Council in 1919 March a Memorandum was read from the Geophysical Committee of the British Association with reference to the R.A.S. taking over the arrangement of these meetings. They are to be considered additional meetings under Bye-Laws, section VIII A,* and are to be in charge of a Geophysical Committee appointed by the Council. The following Societies were invited to associate themselves with the meetings by making them known to their members and by each proposing one representative for appointment to the Geophysical Committee : R. Geographical, R. Meteorological, Geological, Physical Societies, and the British Astronomical Association. The Royal Astronomical Society now from time to time issues supplementary *Monthly Notices* on geophysical subjects at the discretion of the Council.

Though not directly connected with the Society, we cannot close this short account of its work during the last forty years without greeting with pleasure the birth of a new international organisation for the advancement of Astronomy, in which our Society is officially represented. As the various international scientific associations had become inoperative during the war, conferences of delegates of the leading Academies of the nations at war with the Central Powers resolved in 1918 October and November that new organisations should be established, which representatives of neutral nations might eventually be invited to join. A further conference of delegates was held at Brussels in 1919 July, when the " International Research Council " was constituted. The formation of international unions for Astronomy and Geophysics, decided on at the previous conferences, was completed, and statutes for them were agreed to. In each of the countries participating, a National Committee for the promotion of astronomical work and for nominating delegates to the meetings of the astronomical union is formed, and on the one representing this country the R.A.S. appoints six members. The work of the " Union Astronomique Internationale " is to be carried on through various Committees, each having its own Chairman. Under the statutes, the Chairman and members of these Committees are elected by the General Assembly of Delegates, but each Committee has power to add to its number and to draw up its own regulations, subject to approval by the General

* Addition to the bye-laws passed 1918 June 14 (*M.N.*, **78**, 544).

Assembly. General meetings of the Astronomical Union are as a rule to be held every three years, and the first was held at Rome in 1922.

Every worker in Astronomy will earnestly hope that this international organisation will succeed in its principal purpose, to facilitate the relations between astronomers of different countries where international co-operation may be necessary or useful.

II

When the Society was founded, most people dined at or about five o'clock and had tea about nine o'clock. It was therefore natural that the ordinary meetings were held at eight o'clock, and that the Fellows had an informal meeting with tea afterwards, about ten o'clock. For some (now unknown) reason the Annual General Meeting in February began at three o'clock. The hour of dinner which had been in constant motion since the Middle Ages (when it was noon or 11 a.m.) continued to move on and changed place with the tea hour; but the Astronomical Society stuck to the eight o'clock rule. In the days of stage-coaches country members probably did not find this inconvenient, as they could not in any case get home the same evening. But when railways came and trains multiplied, complaints began to be heard from people living at a distance from town. Early in the eighties the Royal Society changed its hour of meeting to half-past four, and soon after that the Fellows of our Society who were anxious for a change began to make their voices heard. At the Annual General Meeting in 1885, Mr. Sydney Waters proposed that the ordinary meetings should commence at five o'clock. In addition to the convenience this would be to country members it was pointed out that it was impossible to attend both the meetings of our Society and those of the Royal Institution. The opposition maintained that many people were engaged during the day and would not be free as early as five o'clock; and several speakers then and during subsequent discussions said that the faces seen at the Annual Meetings (at three o'clock) were different from those seen at the ordinary meetings at eight o'clock, which was supposed to show that many members would not be able to come in the late afternoon. But the best trump card was considered to be, that the Council would not be able to get through their work before the meeting; and whenever there was also a meeting of the Library Committee, the case would become quite hopeless. As it was also denied by some, that there was a sufficiently general desire for a change to warrant an alteration of the Bye-Laws, it was finally decided to send out post-cards to all the Fellows, asking their opinion. The result of the voting was :—

In favour of 8 o'clock	. .	104
„ „ „ 5 „	. .	125
Neutral . . .	.	149
Preferring some other hour	.	16
		394

At a Special General Meeting, called chiefly for another purpose in the following May, the resolution that the ordinary meetings be held at five o'clock was put and lost.

The question was next brought forward at the Annual Meeting in 1890, as a motion that business should commence at four o'clock. After some discussion it was decided to defer the question for further consideration by the Council. At their meeting in April, the Council resolved not to give an opinion as to the hour of meeting, but they put it on record, that if the ordinary meeting were held not earlier than half-past four, they would have time to finish their own business before it. A Special Meeting was held in the following December to consider a proposal by Mr. Chambers, that for a period of one year the meetings be held at half-past four. As a Bye-Law could obviously not be altered for a limited period, leave was given to withdraw the words " for the period of one year," but even though the hour of half-past four was amended to five o'clock, the altered resolution was rejected. All the same, Mr. Chambers brought it forward again in the following February, when it was again lost, but this time only by four votes (35 to 31).

The motion next came before the Society at the Annual Meeting of 1894, when it was again negatived, although the Council acknowledged that the change would not interfere with their business. The question was then allowed to rest till 1900, when a Fellow sent in a notice in January ; and as it appeared doubtful whether it had been received a month before the Annual Meeting (as required by the Bye-Laws), it was made a motion of the Council, without expressing any opinion on it. The discussion was very full and animated, and an additional argument was drawn from the fact that the British Astronomical Association met in the afternoon and seemed to prosper. From the way in which the motion was brought forward, Mr. Knobel inferred that its object was rather to elucidate the opinion of the Fellows than with the idea of taking a vote on it. He therefore moved the previous question, which eventually was carried. Many Fellows who voted for this were probably influenced by their regard for the incoming President, who had always spoken strongly against the change.

But a year later, in 1901 February, when the Council brought forward the motion that the hour of meeting should be five o'clock, it was carried by a large majority, and this long-debated question was at last disposed of. A natural consequence was, that after the lapse of another year it was resolved that the Annual General Meeting should also commence at five o'clock.*

Looking back now after twenty years, it is difficult to put oneself in the place of the strenuous opponents of the change. The meetings seem to be as well attended as before, and the afternoon tea between half-past four and five o'clock has quite naturally taken the place of the tea at 10 p.m., which was very much out of date. And the fear that the Council would not be able to get through its work by five o'clock, has proved to be groundless. The Council meets generally at three o'clock, or, if there is a press of business, at half-past two, and has always finished by half-past four. And while formerly the Council in February had to meet a week before the Annual Meeting, it is now able to meet on the day of this meeting at the usual hour.†

Another Bye-Law which was felt as a grievance by many Fellows living at a distance, was the one (No. 12) which prevented them from giving their votes at the election of the Council, unless they were present at the Annual Meeting and personally handed in their voting paper. At the meeting of the Council in 1885 March a letter was read, signed by 118 Fellows, asking that the Bye-Laws might be altered so as to allow Fellows to vote by proxy at the Annual Meeting. The Council decided not to recommend this ; but they were of opinion that voting by post might fitly form a subject of discussion at a Special General Meeting. Soon after, eight Fellows sent in a demand for a Special Meeting to consider an addition to Bye-Law 12, to the effect that ballot papers of absent Fellows be accepted if signed, and another addition to Bye-Law 53, that Fellows may vote at Annual Meetings personally or by proxy.

A Special General Meeting was therefore summoned for May 8, after the ordinary meeting. At this Lord Crawford moved that a balloting list " handed to the Scrutineers of the ballot on behalf of any Fellow not present at the meeting, shall be accepted if duly verified by the signature of the absent Fellow." It was, however, pointed out by the President (Dunkin) on behalf of the Council, that it was not desirable that country Fellows should

---

* By an oversight this is not mentioned in the *Monthly Notices*, **52**, 218.

† The R.A.S. dinner club now dines after, instead of before, the meetings, which is also more in accordance with modern customs.

send their lists through another Fellow ; also that it ought not to be signed, but that it should be sent by post under cover of a letter signed by the absent Fellow. After some discussion the addition to Bye-Law 12 was passed by a large majority in the following form : " Any List addressed and posted to the Scrutineers of the Ballot by any Fellow not present at the meeting shall be accepted, if duly verified by the signature of the absent Fellow on the envelope." A second resolution, that Fellows may vote by proxy on the business brought before an Annual or a Special General Meeting, was lost by a small majority, and has not since been brought forward.

Among the Fellows living at a distance and, therefore, unable to give their votes personally, there has always been a small number of foreigners residing outside the United Kingdom. In 1894 May the President ruled that foreign Fellows are eligible as Associates, and that they would not be debarred from any right which they held by virtue of their Fellowship. Six months later he stated that he had obtained an informal legal opinion as to foreigners being elected Fellows, to the effect that in view of its having been the custom of the Society for a considerable time, they could not now be considered ineligible. In this the Council concurred. A curious consequence of having foreign Fellows is, that a foreign astronomer of distinction, who is already a Fellow, is occasionally elected an Associate, and yet (if he has compounded) continues to remain a Fellow.*

Another question in which all Fellows were equally interested was that of the composition fee. Since 1831 a Fellow who had paid his admission fee of two guineas might compound for his annual contributions by a payment of twenty guineas ; and no matter how many annual payments he had made, he had still to pay twenty guineas if he wished to compound. Though it was manifestly unfair, this arrangement remained in force for over seventy years. In 1895 the Treasurer announced to the Council that he was going to propose the raising of the composition fee to thirty guineas, with the proviso that a Fellow who had paid thirty-five annual contributions should be considered to have compounded. But he withdrew the proposal before it was laid before a meeting. Two years later a distinguished astronomer resigned his Fellowship after paying contributions for about forty years. Even this did not for a couple of years produce any effect, until the Treasurer in 1902 was directed to draw up a memorandum on the subject. This resulted in the following

* On the other hand Brünnow, an Associate, was in 1865 appointed Royal Astronomer of Ireland, and had (in 1869) to be elected a Fellow, while his name was omitted from the list of Associates.

alteration of Bye-Law 23, which was proposed by the Council and passed at the Annual Meeting in 1903 :—

> That the composition fee be raised to thirty guineas, and be reduced by one guinea each year after five years to a minimum of five guineas.

While on the subject of the annual contributions, we may mention that formerly the Council had no discretion to deal with cases of Fellows who were three years in arrear with their subscriptions, but had to expel them. It happened, of course, occasionally that a man got into financial difficulties without any fault of his, in which case it was felt to be very hard to have to deal severely with him. It was therefore proposed by the Council, and passed unanimously at the Annual Meeting in 1900, that if in such a case a Fellow should desire to resign his Fellowship, it shall be in the power of the Council to remit the whole or any portion of his arrears by a special vote in which at least two-thirds of those present and voting shall concur.*

During the war many Fellows found great difficulty in paying their subscriptions. It was thought that some discretion should be allowed to the Treasurer in dealing with such cases. At a Special General Meeting, held on 1917 June 8, an addition to Bye-Law 26 was therefore passed. By it, any Fellow who was on active naval or military service might be excused from payment of his annual contribution until the end of the year following that in which peace was declared. After that he could either resume his Fellowship or resign it without any liability for arrears of subscription.

The general upheaval caused by the war brought about a reform in the constitution of the Society, which had been talked of privately for a long time without being formally proposed— the admission of women as Fellows. When the British Astronomical Association was founded in 1890, women were at once admitted as members. But the lead thus given was not taken up, although the subject had been before the Council as early as 1886. In November of that year, Miss Pogson, of the Madras Observatory, was duly nominated for Fellowship by three Fellows. Before ordering her name to be suspended for election, the Council thought it well to obtain Counsel's opinion on the admission of women. Mr. Ranyard reported, that unless it could be shown that a woman could not consistently exercise the rights and perform the duties of a Fellow, the Council could be compelled to allow the name of a woman to be suspended for election. But when a second opinion was called for, it was to the effect, that regard being had to the social habits of the time when the Charter was granted, female

* Accidentally omitted from the report of the meeting, *M.N.*, **60**, 295.

Fellows were not likely to have been in contemplation; and as the masculine pronoun is used throughout, they must be taken as not included.   The Council, however, decided that Miss Pogson's name should be suspended, so as to forward it to the general body of Fellows to be dealt with by them as they thought fit.   In voting adversely there could be nothing personal, as it would be understood to mean that the question should first be considered by a General Meeting.   Miss Pogson's proposers thought so too, and withdrew her name.

Again in 1892 three ladies were proposed for election and the Council adopted the same procedure, but this time the proposers did not shrink from the ballot.   It was taken at the April meeting * and resulted in the three ladies failing to obtain the requisite three-fourths of the votes given.

The matter was now allowed to rest for many years.   A mild plaister was offered to the possibly wounded feelings of the ladies a few months after their rejection.   It was resolved by the Council that the President be authorised to issue cards of admission to the meetings " to such persons as it may be thought desirable to admit," available for one season (November to June);   the President to submit a list, and one-third of all the members of the Council to be sufficient to veto a name.   This system worked very well, and was continued as long as ladies were ineligible for Fellowship.

At last the matter was settled in 1915.   At the Annual General Meeting the Council proposed that the Meeting should approve of the admission of women as Fellows and Associates, and that it should request the Council to take the necessary steps to render their election possible.   It was explained that enquiries had been made at the Crown Office and the Privy Council Office.   A draft petition to the King in Council and a draft Supplemental Charter had been prepared, the forms of which had been approved by the Privy Council Office.   The expense would be about £100.   During the discussion which followed, not a single voice was raised against the proposal.   It was pointed out that a few years only after the foundation of the Society, Caroline Herschel had been elected an Honorary Member ; she was soon followed by Mary Somerville, and later by Miss Sheepshanks (a benefactress of the Society), Lady Huggins, Miss Agnes Clerke, and Miss Cannon.   The resolution was passed by fifty-nine votes to three.   At the meeting in the following November it was announced from the Chair that the Supplementary Charter had been received.   The first lady-candidates for Fellowship were balloted for and duly elected at the meeting in 1916 January.

* After some would-be facetious remarks by a Fellow about getting a piano and a fiddle, dancing through most of the papers, etc.

We have only to record one other attempt to effect a considerable change in the Bye-Laws. This was made on four different occasions by the late Mr. Ranyard, and the object was to stop the presentation of Gold Medals. He first entered the lists at the Annual Meeting in 1886, when he moved that Bye-Laws 71 to 76, referring to the medal, be repealed. This met with no sympathy, and the arguments adduced by Mr. Ranyard were very few and trivial, being mainly that the giving of medals was wrong in principle and that three whole Council meetings were given up to the discussion as to who was to have the medal. This, by the way, is *not* the case, as there is only discussion at one meeting, and then it does not take up the whole time. Only three members voted for the motion.

The next year Mr. Ranyard tried his luck again, but this time he went on a different tack. He merely moved that no medal be given " unless the nominee selected be a foreign astronomer not resident in Great Britain." His principal arguments were: first, that the Society should not give rewards to its own members, but should be " above suspicion, like Cæsar's wife " ; and secondly, that it might do a man a great deal of harm to be always hoping for distinctions and feeling disappointed if he did not get them. In support of this he quoted some anecdotes from Arago's *Biographies*. But he failed completely to carry the meeting with him, and equally unlucky was an amendment proposed, that no member of the Council should receive the medal.

Nothing daunted, Mr. Ranyard came back again in 1888 February with his original motion of 1886, and that time he very nearly succeeded, as twenty-three members voted for and twenty-three against the motion. But the President gave his casting vote against it. After this approach to a success, Mr. Ranyard waited two years before renewing his onslaught. In 1890 February he again moved that no medals should be given to anybody. But the Fellows present were evidently tired of the whole thing, and "the previous question " was carried by a large majority. This was the last time Mr. Ranyard tried to persuade the Society as to this matter, about which he was evidently very much in earnest. He died in 1894, and nobody has ever tried to take up his proposal. The Gold Medal is still given to some astronomer after a full and searching discussion by the Council, and is always accepted with pride and pleasure by the recipient. And nobody has been found to insinuate, even obscurely, that the Council might try and be more like Cæsar's wife.*

* All the same, the Council in 1917 January passed a resolution that it is " undesirable " that a Member of Council be proposed for the Medal, unless the proposal is supported by a written notice sent to the President before the

Soon after the cessation of these attacks on the Gold Medal, the Council was called upon to issue regulations for another prize to be given in the name of the Society. The founder of this prize was Mrs. Hannah Jackson, a daughter of the well-known architect and writer on architecture, Joseph Gwilt.* In 1861 June she announced to the Council of the Society her desire to give £300 new 3 per cent. stock, to the intent that the same might be a reversionary gift to the Royal Astronomical Society, to be held on trust, and the dividends thereof to be given either annually or every two years to " any person writing the best astronomical work or in any other way advancing Astronomy, either by the invention of a new astronomical instrument or by the discovery of any new heavenly body." The gift was to be called the " Hannah Jackson (*née* Gwilt) Gift," and the donor wished to receive the interest of the £300 during her life. At the request of the Council, Mrs. Jackson consented to the limit of the term of the accumulation of the interest being extended to seven years; she also agreed that the gift should consist of a medal, of money, or of both.† The gift was then accepted by the Council and was announced in the Council Report of 1862 February, and by the adoption of this Report the Annual General Meeting was supposed to have adopted the general principle of the award of another medal.

Mrs. Jackson died on 1893 December 1, after which date the interest was allowed to accumulate for a couple of years. When the question of the new medal was considered by a Committee of the Council, it was at first proposed that the medal be of silver and made from the same die as the Gold Medal, to be inscribed " Hannah Jackson (*née* Gwilt) Gift," with name and date. But on second thoughts it was decided that the obverse should show the portrait of William Herschel, and the reverse Urania holding a small armillary sphere; the medal to be in bronze and three inches in diameter. The medal and money grant to be awarded at intervals of not less than three nor more than seven years. As the Committee had dealt with these matters with great deliberation, nearly the whole of the year 1896 had passed before the Society was informed of these arrangements. As it was desired to make the first award of the medal and a cheque in the following February, a Special General Meeting was called for January 8

November Meeting of Council, signed by not less than seven members of Council. But this is merely an expression of opinion, not a bye-law.

 * F.R.A.S., died 1863. His daughter latterly called herself Mrs. Jackson-Gwilt. She was a somewhat eccentric lady, who had a coat-of-arms depicted on her visiting card, over her name. She left a large astronomical scrap-book to the Society's library.

 † Extract from Council Minutes, 1861 June 14, printed in *M.N.*, **57**, 36.

to make a slight alteration of Bye-Laws 69–74, by adding everywhere the word " gold " before " medal." This would have the effect of not forbidding the bestowal of the new medal. At the February meeting in 1897 the first award of the " Hannah Jackson (née Gwilt) Gift and Medal " was made to the veteran astronomer, Lewis Swift, who was well worthy of the honour of heading the list of recipients of this new mark of our Society's appreciation.

It has always been considered something in the nature of a treat to listen to the generally admirable addresses from the Chair before the presentation of the Gold Medal. But the same can hardly be said of the rest of the entertainment provided on these occasions, consisting of scraps of Observatory Reports and one or two obituary notices, read aloud by one of the Secretaries. These may be very interesting to read in print, but are hardly suitable for oral delivery. It was therefore a most welcome innovation when Professor Newall, after the presentation of the medals in 1909, delivered a second address. He said he had gathered that the view was held in many quarters that the President might, from time to time, make the Annual Meeting an opportunity for a brief address on some subject of his own choosing, preferably on some special branch of our subject, compiled from a special point of view, whether retrospective and historically suggestive, or prospective and speculatively suggestive. On this occasion he wished to lay before the meeting the following aspect in astrophysical questions : " Can it be that the main characteristic spectroscopic phenomena of the sun and stars are dictated mainly by matter continually streaming in from without, and not mainly by matter brought from within the body of the sun and the star ? " The thorough discussion of this and associated questions by a speaker who is devoting his life to the study of them, makes this address take a high rank among the many important communications made to the Society in the course of years.*

Equally important (in a different way) was the subject dealt with by Gill two years later before leaving the Chair, when he drew attention to the desirability of enlarging the *Nautical Almanac* office, to enable the Director to devote his time to researches in astrodynamics. We have already alluded to this address.

Since 1911 no President has followed the example thus set of delivering a second address, for the interesting address on the foundation of the Society by Professor Fowler (1920 February) can hardly be classed with those just mentioned. It was delivered on a very special occasion, to which it was altogether devoted, and there was, moreover, no presentation of a medal that year, and, therefore, the usual address could not be given. An address,

* The address is printed in full in the *M.N.*, **69**, 332–344.

reviewing in detail the work of the medallist, must generally be the outcome of a considerable amount of preparation on the part of the President, and it is perhaps not fair that a busy man should be expected to undertake extra work at the same time. But occasionally, when a President has something very much at heart connected with his own work, and yet perhaps not ready to be embodied in a formal paper, a second address would enable him to feel : *liberavi animam meam.*

Special addresses have sometimes been delivered by foreign Associates. Thus in 1897, Professor Barnard crossed the Atlantic to receive the Gold Medal, but was delayed by fog till after the Annual Meeting. An informal extra meeting was therefore held on March 2 following, when Professor Barnard exhibited and explained a number of slides of planets, comets, parts of the Milky Way, etc. Ten years later, 1907 June 26, there was a Special Evening Meeting to hear another distinguished American, Professor Hale, give a lecture on " The Opportunities for Astronomical Work with inexpensive Apparatus." The members of the British Astronomical Association were invited.

At the very end of the period we are here reviewing, on 1919 December 12, the ordinary meeting was not devoted to the reading of papers. In view of the widespread interest in the theory of relativity caused by the publication of the results of the eclipse observations, the Council thought they would meet the wishes of the Fellows by giving up the whole time of the meeting to the consideration of this subject. Addresses were delivered by Professor Eddington and Mr. Jeans, after which Sir Oliver Lodge also addressed the meeting, and a communication from Sir Joseph Larmor was read.* It was a fitting tribute to the great theory, which had been raised from the rank of an interesting hypothesis to that of an epoch-marking theory by the confirmation afforded it by experiment and observation.

Apart from these special occasions all ordinary meetings continue to be devoted to the reading of papers and the subsequent discussion on their contents.† This discussion is generally very interesting and often throws additional light on the subject. It is a great pity that we possess no records whatever of what happened at the meetings previous to 1862 November, when the *Astronomical Register* began to give short reports of the proceedings at the meetings. From that time we possess an unbroken series of accounts of these, as *The Observatory*, from the appearance of its

* These addresses and papers were all printed in the *Monthly Notices.*

† As a rule, there are more papers sent in than can be " read " at a meeting ; but in 1887 December only two papers had been received, and neither of these was suitable for reading. This provided a rare opportunity for discussing various topics of special interest at the moment, such as curved plates, etc.

first number in 1877 April, has made these reports a special feature, to which most readers probably look first of all on receiving a new number. The lantern provided in 1890 December has been a most useful acquisition, as it not only enables the members present to see under the best conditions the wonderful features of the Milky Way, Nebulæ, Solar Corona, etc., from the most recent photographs, but also gives an author convenient means of showing tabular matter or intricate formulæ on the screen instead of wasting time in copying them out on the blackboard, where many of the audience are unable to see them. Another useful innovation was tried in 1889, when advertisements were inserted in *The Times* giving the titles of the papers received up to a couple of days before each meeting. This was discontinued a year later, since when postcards giving this information have been sent to such Fellows as express their wish to receive them.

The principal work of a scientific Society, and that by which posterity will judge it, is, of course, the publication of papers. During the forty years ending 1920 our Society has published a steadily increasing number of papers dealing with all branches of Astronomy. At the beginning of this period, in 1881 November, Cayley resigned the Editorship of the Society's publications, which he had held (in succession to Grant) since 1859 December, except during the two years (1872–74) when he was President, while Proctor took his place as Editor. It was resolved by the Council in 1881 December, that in future the publications should be edited by the Secretaries, with such help from the Assistant Secretary as they might require, and that the remuneration of £50 per annum hitherto paid to the Editor, be paid to him. This arrangement is still in operation and has worked well.

It is very curious to see how the *Monthly Notices* have gradually taken the place of the *Memoirs* as the principal organ of the Society. This was, of course, due to the rapidity and regularity of its publication. Though this journal had gradually taken over many short papers from the *Memoirs*, the Annual Report appeared in both series up to 1858, while each volume of the *Memoirs* was described on the title-page as the quarto half-volume for the session . . ., and bore the following notice on the cover : " The Octavo Half-volume, being volume . . . of the *Monthly Notices*, containing the abstracts, observations, shorter papers, etc., for the same session, is given to purchasers of the Quarto Half-volume, and is necessary to complete it." But this notice disappeared after 1858, the *Monthly Notices* being thus recognised as a separate journal. Still the Council wished to make the journal " become an integral part of the volumes of the *Memoirs*." This was done by re-imposing the type into a quarto form with double columns, thus forming an edition of the *Monthly Notices* which might be bound up with the *Memoirs*. In this way volumes **19** to **27** appeared in a double form, after which this curious arrangement was discontinued. It was indeed totally unnecessary, as the *Monthly Notices* had taken its place in the first rank of astronomical publications. It has more than once been pointed out by the Council, that the Society does not print papers which have already appeared elsewhere in print ; but foreign astronomers not infrequently make the *Monthly Notices* the medium of publication of their work. It is interesting to note the increase in the number of pages per volume during late years :—

Vol.		pages	Vol.		pages
Vol. 25	1864–65	283 pages	Vol. 55	1894–95	553 pages
,, 30	1869–70	231 ,,	,, 60	1899–1900	632 ,,
,, 35	1874–75	416 ,,	,, 65	1904–05	893 ,,
,, 40	1879–80	637 ,,	,, 70	1909–10	680 ,,
,, 45	1884–85	525 ,,	,, 75	1914–15	727 ,,
,, 50 *	1889–90	568 ,,	,, 80	1919–20	820 ,,

An important part of each volume is the Annual Report of the Council. This has always been a very readable document, more so perhaps during the first thirty or forty years, when the Council or the Secretaries told their story in their own words. But whatever the Report has lost since then as literature, it has certainly gained as a scientific record, particularly as regards the " Notes on some points connected with the recent progress of Astronomy." These formerly confined themselves to notices on a limited number of important works and Memoirs. But from 1893 an attempt has been made to give a list of discoveries during the year, not only in the cases of minor planets and comets, but in those of double stars, variable stars, stellar spectra, etc. The copious references to the recent literature on these subjects must be a great boon to workers, as it is now more than ever exceedingly easy to overlook some paper or note among the vast multitude published, some of them in journals where one might not think of looking for them.

The value of the *Monthly Notices* as a continuous record of astronomical progress is illustrated by the occasional requests from abroad for copies of those old volumes of the journal which have been out of print for many years. We have mentioned † that volumes **3, 4, 5** have been unobtainable almost from the day they were printed. Volumes **7** and **27** have also been out of print for more than forty years, while there are only a few copies left of several others. Enquiries were made in 1911 as

---

* In addition to an appendix of 175 pp. See above, p. 220.
† Above, p. 80.

to the cost of reprinting **3, 4,** and part of **5,** but even then it was prohibitive. Of the papers published in these old books very few are now of great importance, but everyone fond of books hates to have gaps in a long series. One paper was, however, of first-rate importance, that of Adams on the Orbit of the November Meteors in volume **27.** In view of the expected return of the Leonids, this paper was in 1897 reprinted in volume **57,** and that is probably all that will ever be reprinted. It is much more unfortunate, that during a destructive fire at Messrs. Neill & Co.'s, the Society's printers, in 1916 May, many copies of recent numbers of the *Monthly Notices* were lost, so that there are very few copies left, particularly of volume **76,** numbers 1–5.

Whether a paper is to be printed in the *Monthly Notices* or in the *Memoirs* depends now almost altogether on whether the extent of tabular matter contained in it requires the quarto size or not. A troublesome relic of old times was a rule requiring a Fellow to call in person for his copy of the *Memoirs* or to get somebody else to call for it. Naturally people often forgot to do so within the prescribed limit of time and had to make special application to the Council before they could get their copies. The introduction of parcel post facilitated the abolition of this tiresome rule ; from 1891 the Society has paid the postage on the *Memoirs*, and from 1903 it has been unnecessary to apply for each volume. A more serious cause of complaint was the long delay which often occurred in publishing a paper in the *Memoirs*, before a number of papers sufficient to fill a volume had been collected. As an experiment, some copies of volume **47** were printed in six parts, but there was not much demand for them, and the experiment was not repeated for more then twenty years. Since 1905, however, beginning with volume **57,** each paper has been printed and distributed separately with as little delay as possible.

It may be said without fear of contradiction that the volumes of *Memoirs* published during the last fifty years are not inferior in value to the earlier volumes. And they have one advantage over those. Not one of the later volumes is partly filled with longitude determinations or results of meridian observations, which though generally printed without expense to the Society, should not have been inserted in the *Memoirs*.

We have already mentioned when describing the Society's participation in eclipse expeditions, that arrangements were made with the Royal Society in 1898 whereby a sufficient number of copies of the eclipse reports printed in the *Proceedings of the Royal Society* and *Philosophical Transactions*, were supplied to our Society to be re-issued as appendices to the *Monthly Notices* and *Memoirs*. This arrangement was very soon extended so as

16

to include papers on astrophysical subjects not connected with eclipses. In this way appendices with double pagination were issued to volumes **58** and **60** to **65** of the *Monthly Notices.* Though the Fellows were thus supplied with copies of valuable papers which many of them would not otherwise have seen, this arrangement had two drawbacks. It increased the thickness of the volumes, which even without them was rather considerable, and it prevented the index being at the end of the volume. It came to an end when the size of the *Proceedings* was changed from demy 8vo to royal 8vo.

Appendices of similar contents appeared to volumes **54, 55, 57** of the *Memoirs.* But in 1906 the Council of the Royal Society discontinued the agreement, apparently because they had received many applications from other Societies for the privilege of including Royal Society papers in their publications. They declared themselves ready to consider the question of supplying either series A or B of the *Philosophical Transactions* direct to members of Societies at a reduced rate, " provided such Society be willing to subscribe on behalf of an adequate proportion of its members." But our Council did not accept this invitation. In 1909 March the Council resolved that the Secretaries be encouraged to insert in the *Monthly Notices* brief abstracts of papers in the *Philosophical Transactions* and *Proceedings of the Royal Society.* The encouragement thus given appears to have been insufficient, and the short reviews often found in the *Monthly Notices* sixty to seventy years ago have never been resumed.

Apart from its regularly appearing publications, the Society has only on one occasion shared in the printing of an astronomical work. This was in 1910–1912, when the Society joined the Royal Society in publishing " The Scientific Papers of Sir William Herschel," in two large quarto volumes. It had always been felt as a serious desideratum in astronomical literature, that the important papers of W. Herschel had never been collected, but had to be looked for in about forty volumes of the *Philosophical Transactions.** But the difficulty of getting a private publisher to undertake the risk of issuing so extensive a work was so great, that nothing was done to realise the no doubt widely spread wish till nearly ninety years after Herschel's death. When an appeal had been made in print, addressed to the R.A.S. and the Royal Society, by Professor See of California,† the matter was at length taken up in earnest in the beginning of 1910. A joint Committee was formed, and thanks to the liberality of Sir W. J. Herschel, the grandson and son of two great astronomers, access was given

---

* See, for instance, W. Struve, *Études d'astronomie stellaire*, p. 23.
† *The Observatory*, **32**, 473.

to the manuscript treasures at that time still preserved at Slough. The Royal Society also lent Herschel's original " Sweep-Books," so that it became possible to make a thorough revision of his three catalogues of Nebulæ and Clusters. A lengthy biographical introduction was prepared from letters and other manuscripts, and the papers read before the Philosophical Society of Bath were now printed for the first time. The new edition came out in the spring of 1912.*

Without having any pecuniary interest in the undertaking or assuming any editorial responsibility, the Society lent a hand in the republication and distribution of Franklin-Adams' *Chart of the Heavens* and Higgs' smaller atlas of the *Solar Spectrum*, in 1913 and 1915.

In addition to receiving the *Memoirs* and *Monthly Notices*, attending the meetings and voting on matters brought before the General Meetings, the Fellows have the privilege of using the fine library gradually collected by the Society. The greater part of this is contained in the large room on the first floor, in which the Fellows assemble before the meetings. This was in 1890 connected with the gallery at the top of the room by means of a spiral staircase. At the same time the Society was put to considerable expense in carrying out various structural alterations made necessary by the rapid increase of the library. The Treasury declined to let these be done at the public expense, giving as a reason that it had originally been intended that the library should be of the full height of the first and second floors together and provided with galleries. But in the building as constructed, when handed over to the Society, an intermediate floor was introduced and a depository fitted with shelving for large parcels on the second floor. As the building had all the appearance of having been permanently completed for the purpose for which it was used, the Treasury could not sanction any further work on it at the public expense. The Society therefore undertook to convert the room above the library (which has the same floor-space but is much lower) into an additional library. This " Upper Library " contains books not dealing specially with astronomy, such as the Transactions of various Foreign and Colonial Societies, duplicates, etc. Some of these duplicates are of great value and have been acquired for the express purpose of being lent to Fellows, without disturbing the sets in the library, which are never lent out, such as the *Astronomische Nachrichten*, and our *Memoirs* and *Monthly Notices*.

* When Herschel's MSS. were presented to the Society in 1918 January, it was found that a summary of his observations of variable stars compiled by Caroline Herschel had not been at the disposal of the Editor of the *Scientific Papers*. It was printed in the *M.N.*, **78**, 554–568.

The problem of how to find room for the ever-increasing additions to the library has of late years repeatedly occupied the attention of the Council. In 1889 and in 1912 some non-astronomical books were sold, but this remedy can obviously only be applied to a very limited extent. It has, however, been ascertained that the floor of the upper library would be strong enough to support an additional back-to-back bookcase down the centre of the room if not more than 7 feet high. The question of shelf-space is therefore not of immediate urgency.

A catalogue of the library, complete to 1884 June and extending to 408 pp. 8vo, was published in 1886. A supplement to 1898 June was brought out in 1900. Since then only the annual lists of accessions, separately paged, have been issued with the last number of every volume of the *Monthly Notices*, and for the convenience of those who do not wish to bind them with this journal, title-pages to these lists were supplied in 1905 and 1912. It is very inconvenient not to have a catalogue of this valuable collection complete to a recent date, but the great increase in the cost of printing will make it very difficult to supply this want.

The manuscript department of the library, though only possessing a few *old* manuscripts, is nevertheless an important part of the Society's property. Some important additions to it have been presented within the last few years, the most valuable one being the great collection of William Herschel manuscripts, presented by the late Sir W. J. Herschel. A detailed descriptive catalogue of this collection was given in the *Monthly Notices*, volume 78.

In connection with the library, we may also mention that the Society possesses a fine collection of portraits. Round the walls of the meeting-room are arranged framed photographs of all the past Presidents, while there are also paintings of Newton, W. Herschel (a copy of Artaud's portrait of 1819), and Goodricke. The fine portrait of Baily still hangs in the Council-room.

Finally, it may not be out of place to say a few words about the Council, the governing body of the Society. Consisting originally of seventeen members, of whom eight did not hold any special office,* these numbers were in 1825 raised to nineteen and ten. They remained unaltered till 1858, when advantage was taken of a clause in the Charter permitting the addition of two members to those who hold no office. The two principal classes of Fellows constituting the Society, professional astronomers and amateurs, are always both well represented, and in particular it is an unwritten rule that of the two Secretaries one is a professional and the other an amateur, or at least that one is specially con-

* In 1920 there were only six of these.

versant with gravitational Astronomy, the other with observational Astronomy or astrophysics. The continuity of the Council has been kept up by always retaining some of the more prominent members for a number of years. We shall only mention a few examples of *continuous* service of Fellows not now living :—

Baily	.	.	.	25 years	1820–44
E. Riddle	.	.	.	27 ,,	1825–52
Lee	.	.	.	38 ,,	1829–67
Airy	.	.	.	56 ,,	1830–86
Cayley	.	.	.	35 ,,	1858–93
Adams	.	.	.	32 ,,	1860–92
Huggins	.	.	.	46 ,,	1864–1910
Dunkin	.	.	.	23 ,,	1868–91

In addition to these there were several others who served many years on the Council, such as De Morgan, 1830–62, except the year 1845 (and he was Secretary 1831–39 and 1847–55), and Christie, 1872–1913, except 1879–80. The case of Huggins is altogether unique, because he held office for so many years. He was Secretary 1867–72, Foreign Secretary 1873–76, President 1876–78, and again Foreign Secretary 1883–1910, till the day of his death.

While care is thus taken that experience and knowledge of the Society's affairs are not lost, stagnation is prevented by two important provisions in the Bye-Laws. First, that no Fellow who has been President or Vice-President for two successive years shall again be eligible to the same situation till the expiration of one year from the termination of his office. And secondly, that eight only of the twelve members of the Council holding no office, who have served during any year, shall be eligible in the same capacity for the ensuing year.

The gradual increase in the number of Fellows may be seen from the following table :—

	Fellows.	Associates.
1830	243	34
1840	307	38
1850	349	57
1860	380	51
1870	509	45
1880	591	43
1890	595	48
1900	635	48
1910	682	47
1920	715	50

There is therefore every reason to hope that the number of

Fellows will not fall off, particularly as ladies have now at last been admitted.

If this review of the history of the Society had been written six or seven years earlier, we should have been able to look with confidence to the future, expecting the steady financial prosperity, which we have hitherto enjoyed, to continue. But the cost of printing has increased enormously during the last few years, and it does not seem at all likely that it will ever be materially reduced. Still, there is no cause for immediate anxiety. In the past, generous benefactors have occasionally by bequest increased the funds of the Society. Thus, during the last forty years we have received :—

The McClean bequest of £2000, free of legacy duty, from Frank McClean, LL.D., F.R.S., in 1905.

The Farrar bequest of £100, free of legacy duty, from the Rev. A. S. Farrar, D.D., in 1906.

The Gill bequest of £250, free of legacy duty, from Sir David Gill, F.R.S., in 1919, to be devoted to the completion of some great piece of work. It was spent on printing Professor Sampson's Theory of Jupiter's great satellites (*Memoirs*, **63**).

It is surely permitted to hope that the Society will also in future, from time to time, see its funded property increased by similar donations, which will be the more acceptable, the less trammelled by conditions they are.* As regards printing, all scientific Societies in the world seem to be affected in the same way as we are, and many of them are probably worse off, having much less invested capital.

In conclusion, let us glance at what the Society has accomplished during the hundred years it has been in existence, and ask whether this can compare favourably with what the founders of the Society declared to be their objects.

In the Address circulated before the first public meeting the original members summarised the means by which they proposed to advance Astronomy as follows : Collecting, reducing, and publishing useful observations and tables ; setting on foot a minute and systematic examination of the heavens ; establishing communications with foreign observers, circulating notices of remarkable phenomena about to happen and of discoveries ; proposing prize-questions and bestowing rewards on successful research.

---

* Since this was written, the Society has, in 1922 April, received a liberal donation of £2500 from the Hon. Sir Charles Parsons, K.C.B., as a Memorial to his father, William, third Earl of Rosse, the maker of the great telescope. The Society also, about the same time, received a number of smaller donations, amounting in all to nearly £1400, thanks to the energy of the Treasurer, Colonel Grove-Hills. See also the Preface as to his bequest of a Library.

COLONEL E. H. GROVE-HILLS
(1864–1922)
TREASURER, 1905–13 AND 1921–2 ; PRESIDENT, 1913–15.

To face p. 246]

That the publications of the Society have been the means of making a great many observations public property is an undoubted fact.   But the fulfilment of the second promise was never even attempted.   It included "the formation of a complete catalogue of stars and of other bodies, upon a scale infinitely more extensive than any that has yet been undertaken, and that shall comprehend the most minute objects visible in good astronomical telescopes."

Five years after the issue of this manifesto, Bessel invited the co-operation of observers to construct star maps of the region between ±15° Declination.   This was only a small portion of the heavens, yet the work took thirty-four years to complete.   On the other hand, when Argelander took in hand the construction of similar maps of the whole northern hemisphere, based solely on new observations, he and two devoted assistants completed their great task in eleven years, including the printing of a great atlas and a catalogue in three volumes of 324,000 approximate star places.   The desire was at once expressed that this work might be continued to the South Pole, and several southern observatories were to divide it between them.   But nothing came of this project ; while Schönfeld alone in a few years continued the work to −23° Declination.   Again, the great undertaking of the *Astronomische Gesellschaft*, the catalogue of all stars down to the ninth magnitude from zone observations, took for the northern hemisphere more than forty years to finish, chiefly because

" On the strength of one link in the cable
Dependeth the might of the chain,"

and there is generally more than one faulty link in co-operative chains.   This has also been the case with the photographic chart of the heavens, while Gill and Kapteyn rapidly carried out their photographic continuation of Argelander's and Schönfeld's work.

There seems, therefore, no reason to regret that our Society has never wasted time and energy in organising co-operative undertakings.   Again and again history has shown that what is wanted for a successful undertaking in practical Astronomy, as in every other great and laborious undertaking, is the proper kind of man for the work.   If he be found and be given the material means necessary for the realisation of his ideas, he will carry out the work far more quickly and perhaps better than a dozen people acting under instructions from a central institution could complete their tasks.   And this is not only the case with charts and catalogues of stars.   The planetary tables of Le Verrier and

Newcomb, the tables of the moon of Hansen and Brown, and many other results of lengthy researches have been carried out without splitting up the work among a number of collaborators in different places.

But even if it be admitted that co-operation is desirable in many cases, it is by international associations that it should be organised, and not by a Society which, like ours, has a national character, even though it does not exclude foreigners from the ranks of its members. The principal aims of the Royal Astronomical Society were therefore from the beginning, first, the publication of papers ; secondly, to encourage the study of Astronomy in this country and to guide it into proper channels, so as to produce work fruitful for the progress of astronomical science. For forty years previous to 1820, William Herschel had almost every year published one or more papers containing remarkable discoveries or brilliant ideas on the construction of the Universe. But his voice was that of one crying in the wilderness. His son wrote long afterwards of this period : " Mathematics were at the last gasp and Astronomy nearly so ; I mean in those members of its frame which depend upon precise measurement and systematic calculation." That our Society contributed greatly to change this state of things in the course of the first ten or fifteen years of its existence, cannot be disputed. The gradual formation of a steadily growing class of amateur astronomers, the reform of the *Nautical Almanac*, the new spirit infused into the public observatories, the valuable papers published in the *Memoirs*, showed very soon that the new Astronomical Society was successfully endeavouring to advance science. And throughout the century the Society has remained true to the ideals which animated its founders. It may certainly claim a fair share of the credit of the widely spread interest in Astronomy which was manifested by the foundation of the British Astronomical Association in 1890. On that occasion our Society did not copy the dog-in-the-manger policy of Sir Joseph Banks in 1820, but saw with pleasure the rise of another organisation binding together British amateur astronomers and others interested in Astronomy.

The most conspicuous service which the Society has rendered to science is, however, the ready means it has offered for the publication of astronomical papers, lengthy as well as short ones. If the Society had not existed, these would have been scattered, and most of the longer ones would probably have found their way to the publications of provincial Societies and would not have been readily accessible to many students. Even under existing circumstances many astronomers prefer to send their most important papers to the Royal Society or to some local Society, an inconvenience which no bibliographical indexes or reviews yet devised have been able to mitigate. But this is no fault of our Society, and future historians of science will readily acknowledge that it has been mindful of its motto—

Quicquid nitet notandum.

APPENDIX.

I. PRESIDENTS OF THE SOCIETY.

1821–23.	Sir William Herschel.		1872–74.	Arthur Cayley.
1823–25.	H. T. Colebrooke.		1874–76.	J. C. Adams.
1825–27.	Francis Baily.		1876–78.	William Huggins.
1827–29.	J. F. W. Herschel.		1878–80.	Lord Lindsay.
1829–31.	Sir James South.		1880–82.	J. R. Hind.
1831–33.	Bishop Brinkley.		1882–84.	E. J. Stone.
1833–35.	Francis Baily.		1884–86.	Edwin Dunkin.
1835–37.	G. B. Airy.		1886–88.	J. W. L. Glaisher.
1837–39.	Francis Baily.		1888–90.	W. H. M. Christie.
1839–41.	Sir J. F. W. Herschel.		1890–92.	J. F. Tennant.
1841–43.	Lord Wrottesley.		1892–93.	E. B. Knobel.
1843–45.	Francis Baily.		1893–95.	W. de W. Abney.
1845–47.	W. H. Smyth.		1895–97.	A. A. Common.
1847–49.	Sir J. F. W. Herschel.		1897–99.	Sir R. S. Ball.
1849–51.	G. B. Airy.		1899–00.	Sir G. H. Darwin.
1851–53.	J. C. Adams.		1900–01.	E. B. Knobel.
1853–55.	G. B. Airy.		1901–03.	J. W. L. Glaisher.
1855–57.	M. J. Johnson.		1903–05.	H. H. Turner.
1857–59.	George Bishop.		1905–07.	W. H. Maw.
1859–61.	Robert Main.		1907–09.	H. F. Newall.
1861–63.	John Lee.		1909–11.	Sir David Gill.
1863–64.	G. B. Airy.		1911–13.	Sir F. W. Dyson.
1864–66.	Warren De la Rue.		1913–15.	E. H. Hills.
1866–68.	Charles Pritchard.		1915–17.	R. A. Sampson.
1868–70.	R. H. Manners.		1917–19.	P. A. MacMahon.
1870–72.	William Lassell.		1919–21.	A. Fowler.

II. TREASURERS.

W. Pearson	. . 1820–31	A. A. Common	. . 1884–95
J. Lee	. . 1831–40	E. B. Knobel	. . 1895–00
G. Bishop	. . 1840–57	W. H. Maw	. . 1900–05
S. C. Whitbread	. . 1857–78	E. H. Hills	. . 1905–13
F. Barrow	. . 1878–84	E. B. Knobel	. . 1913–22

III. SECRETARIES.

C. Babbage	.	. 1820–24	E. J. Stone	. . 1866–71
F. Baily	.	. 1820–23	W. Huggins	. . 1867–72
J. Millington	.	. 1823–26	E. Dunkin .	. . 1871–77
O. G. Gregory	.	. 1824–28	R. A. Proctor	. . 1872–74
W. S. Stratford	.	. 1826–31	A. C. Ranyard	. . 1874–80
R. Sheepshanks	.	. 1828–31	J. W. L. Glaisher	. 1877–84
A. De Morgan	.	. 1831–39	W. H. M. Christie	. 1880–82
J. Wrottesley	.	. 1831–33	E. B. Knobel	. . 1882–92
G. Bishop	.	. 1833–39	G. L. Tupman	. . 1884–89
T. Galloway	.	. 1839–41	A. M. W. Downing	. 1889–92
H. Raper	.	. 1839–42	E. W. Maunder	. . 1892–97
R. Main	.	. 1841–46	H. H. Turner	. . 1892–99
R. W. Rothman	.	. 1842–43	H. F. Newall	. . 1897–01
T. Galloway	.	. 1843–45	F. W. Dyson	. . 1899–05
W. Rutherford	.	. 1845–47	E. T. Whittaker	. . 1901–07
R. Sheepshanks	.	. 1846–47	T. Lewis	. . 1905–09
T. Galloway	.	. 1847–48	S. A. Saunder	. . 1907–12
A. De Morgan	.	. 1847–55	A. R. Hinks	. . 1909–13
R. H. Manners	.	. 1848–57	A. S. Eddington	. . 1912–17
W. De la Rue	.	. 1855–63	A. Fowler	. . 1913–19
R. C. Carrington	.	. 1857–62	A. C. D. Crommelin	. 1917–23
C. Pritchard	.	. 1862–66	T. E. R. Phillips	. . 1919–
R. Hodgson	.	. 1863–67		

IV. FOREIGN SECRETARIES.

J. F. W. Herschel	.	. 1820–27	A. Strange .	. . 1868–73
C. Babbage	.	. 1827–29	W. Huggins	. . 1873–76
W. H. Smyth	.	. 1829–40	Lord Lindsay	. . 1876–78
R. W. Rothman	.	. 1840–42	J. R. Hind .	. . 1878–80
T. Galloway	.	. 1842–43	Earl of Crawford .	. . 1880–83 ·
W. H. Smyth	.	. 1843–45	Sir W. Huggins	. 1883–1910
R. Sheepshanks	.	. 1845–46	Sir D. Gill .	. . 1911–14
Sir J. F. W. Herschel	.	. 1846–47	A. Schuster	. . 1914–19
J. R. Hind .	.	. 1847–58	H. H. Turner	. . 1919–
R. H. Manners	.	. 1858–68		

V. ASSISTANT SECRETARIES.

J. Epps	.	. December 1830–March 1838
J. Hartnup	.	. March 1838–November 1843
R. Harris	.	. November 1843–December 1845
J. Williams	.	. April 1846–December 1874
W. H. Wesley	.	. February 1875–October 1922

VI. LIST OF PERSONS TO WHOM THE MEDALS OR TESTI-MONIALS OF THE SOCIETY HAVE BEEN ADJUDGED.

(The Gold Medal is in every case intended except where otherwise stated.)

1824. Charles Babbage.
J. F. Encke.
Charles Rümker (*the Silver Medal*).
J. L. Pons(*the Silver Medal.*)
1826. J. F. W. Herschel.
James South.
Wilhelm Struve.
1827. Francis Baily.
W. S. Stratford (*the Silver Medal*).
Colonel Mark Beaufoy (*the Silver Medal*).
1828. Sir T. Makdougall Brisbane.
James Dunlop.
Caroline Herschel.
1829. William Pearson.
F. W. Bessel.
H. C. Schumacher.
1830. William Richardson.
J. F. Encke.
1831. Captain H. Kater.
Baron Damoiseau.
1833. G. B. Airy.
1835. Lieut. M. J. Johnson.
1836. Sir J. F. W. Herschel.
1837. O. A. Rosenberger.
1839. Hon. John Wrottesley.
1840. Jean Plana.
1841. F. W. Bessel.
1842. P. A. Hansen.
1843. Francis Baily.
1845. Captain W. H. Smyth.
1846. G. B. Airy.
1848. G. B. Airy (*Testimonial*).
J. C. Adams „
F. W. Argelander „
George Bishop „
Lt.-Col. George Everest „
Sir J. F. W. Herschel „
P. A. Hansen „
K. L. Hencke „

1848. J. R. Hind (*Testimonial*).
U. J. J. Le Verrier „
Sir J. W. Lubbock „
Maximilian Weisse „
1849. William Lassell.
1850. Otto Struve.
1851. Annibale de Gasparis.
1852. C. A. F. Peters.
1853. J. R. Hind.
1854. Charles Rümker.
1855. W. R. Dawes.
1856. Robert Grant.
1857. Heinrich Schwabe.
1858. Robert Main.
1859. R. C. Carrington.
1860. P. A. Hansen.
1861. Hermann Goldschmidt.
1862. Warren De la Rue.
1863. F. W. Argelander.
1865. G. P. Bond.
1866. J. C. Adams.
1867. William Huggins.
W. A. Miller.
1868. U. J. J. Le Verrier.
1869. E. J. Stone.
1870. Charles Delaunay.
1872. G. V. Schiaparelli.
1874. Simon Newcomb.
1875. H. L. D'Arrest.
1876. U. J. J. Le Verrier.
1878. Baron Dembowski.
1879. Asaph Hall.
1881. Axel Möller.
1882. David Gill.
1883. B. A. Gould.
1884. A. A. Common.
1885. William Huggins.
1886. E. C. Pickering.
Charles Pritchard.
1887. G. W. Hill.
1888. Arthur Auwers.
1889. Maurice Loewy.

1892.	G. H. Darwin.	1906.	W. W. Campbell.
1893.	H. C. Vogel.	1907.	E. W. Brown.
1894.	S. W. Burnham.	1908.	Sir David Gill.
1895.	Isaac Roberts.	1909.	Oskar Backlund.
1896.	S. C. Chandler.	1910.	Friedrich Küstner.
1897.	E. E. Barnard.	1911.	P. H. Cowell.
1898.	W. F. Denning.	1912.	A. R. Hinks.
1899.	Frank McClean.	1913.	H. A. Deslandres.
1900.	Henri Poincaré.	1914.	Max Wolf.
1901.	E. C. Pickering.	1915.	A. Fowler.
1902.	J. C. Kapteyn.	1916.	J. L. E. Dreyer.
1903.	Hermann Struve.	1917.	W. S. Adams.
1904.	G. E. Hale.	1918.	J. Evershed.
1905.	Lewis Boss.	1919.	G. Bigourdan.

THE HANNAH JACKSON (*NÉE* GWILT) GIFT AND MEDAL.

1897.	Lewis Swift.	1909.	P. J. Melotte.
1902.	T. D. Anderson.	1913.	T. H. E. C. Espin.
1905.	John Tebbutt.	1918.	T. E. R. Phillips.

INDEX.

Acceleration, lunar, 158.

Activity of Fellows, 85, 86.

Adams, discovery of Neptune, 92–97; testimonal, 99; orbit of Leonids, 162; address on presenting medal to Le Verrier, 162.

Address on foundation of Society, 3–6.

Addresses, presidential, 237.

Advowsons, 81, 83, 124, 200.

Airy on South's telescope, 54; reform of *Nautical Almanac*, 60; work at Cambridge, 68, 71; position of Ecliptic, 70; Astronomer Royal, 71; medal for Greenwich planetary reductions, 92; testimonial, 93, 99; his attitude to Adams and Le Verrier, 93; defence of himself, 96; his great activity, 148; transit of Venus, 180 *sqq.*; on Smyth's Bedford Catalogue, 202 *sqq.*; on bestowing medal, 206; on endowment of research, 210.

Amateurs in the thirties, 77; in the seventies, 188.

American astronomers, 104; Associates, 105.

Apartments, 35, 63; at Somerset House, 64, 82; offer of, at Kensington, 121; move to Burlington House, 186.

Appendices to *M.N.* and *Memoirs*, 241.

Argelander, testimonial, 99; *Durchmusterung*, 116, 144, 247.

Armagh, work at, 72, 73.

Arrears of subscriptions, 81, 126, 233.

Assistant Secretary, Epps, 63; Hartnup, 63, 84; resident, 82; Harris, 84; Williams, 84, 186; Wesley, 187.

Associates, earliest, 47, 48; American, 105; number and qualifications, 123.

Association, British Astronomical, 248.

Astræa discovered, 92.

Astrophysics, proposed observatory for researches on, 175.

Atkinson on refraction, 70.

Australian solar observatory, 222.

Babbage, founder, 1–3; medal, 43.

Bache, Associate, 105.

Baille, density of earth, 118.

Baily, Ann Louisa, 32.

Baily, Arthur, 2, 32.

Baily, Francis, founder, 2, 3, 20, 22; catalogue of stars, 15; star constants, 15; on Barrett's annuity

tables, 27; tour in America, 29–32; reform of *Nautical Almanac*, 56–61; pendulum observations, 68; standard scale, 69; central figure of R.A.S., 77; on Flamsteed, 78; edits ancient catalogues, 78; pays for *Memoirs*, 78, 83; his portrait, 83; death, 87; his work, 87; medal to, 91.

Baily's beads, 78.

Banks, Sir Joseph, disapproves of the Society, 7.

Barclay, his observatory, 189.

Barnard, extra meeting for, 238.

Barrett, annuity tables, 26.

Beaufoy, medal, 43; instruments presented, 45; observations of Jupiter's satellites, 57.

Becket, Edmund, on transit of Venus, 181; endowment of research, 208.

Bedford Catalogue, 92, 202.

Bedford Observatory, 76.

Bequests, 191, 192, 246.

Bessel, stellar parallax, 70, 89; perturbations of Uranus, 94.

Billycock Hat, origin of name, 25.

Binary stars, orbits of, 75.

Bishop, his observatory, 77, 116; testimonial, 99; President, 141.

Blagg, Miss, lunar nomenclature, 223.

Bond, G. P., discovers Hyperion, 98; photography, 113; Donati's Comet, 157.

Bond, W. C., Associate, 105.

Bowler hat, origin of name, 25.

Boys, density of earth, 118.

Brewster, life of Newton, 78; address to British Association in 1850, 112, 120.

Brinkley, President, 51.

Brisbane, founder of Paramatta Observatory, 66.

Buckingham, his 21-inch refractor, 189.

Bunsen, solar spectrum, 114.

Burlington House, offered in 1854, 121; move to, 186.

Burnham on Smyth's Bedford Catalogue, 203, 205.

Cagnoli, paper on figure of earth, 20, 56.

Cambridge Observatory, Airy's work at, 68, 71.

Cape, Mural Circle, 73; Observatory, 66, 106; proposed Board of Visitors, 221.

Carnot, moving power of heat, 111.

Carrington, sun-spot observations, 156; observatory, 188; bequest, 191; MSS. presented, 192.
Carte du Ciel, 215.
Catalogue of astronomical literature, 226.
Catalogue of Stars of Society, 15.
Cavendish experiment, 70.
Cayley, editor, 133, 239.
Challis, failed to find Neptune, 64.
Chambers, G. F., 169, 205.
Charter, 50; modified in 1915, 234.
Chevallier, sun-spot observations presented, 193.
Clock Star tables, 14.
Colby, founder, 2, 3; Ordnance Survey, 117.
Colebrooke, founder, 2, 3; Indian scholar, 18.
Comet, Donati's, 113.
Commission, Royal, on advancement of science, 174, 207.
Common on endowment of research, 209; celestial photography, 213.
Composition Fee, 232.
Conjoint Board, 227.
Consort, Prince, death of, 140.
Cornu, density of earth, 118.
Corona, types of, 199.
Council, first, 7, 8; reports, 134, 240; mode of electing, 140; dissensions in, 176 sqq.; long-continued service on, 245.
Cowell, Superintendent of Nautical Almanac, 218.

Davy, Humphry, 37.
Dawes, observations of double stars, 75; medal, 123.
Day, commencement of astronomical, 225.
Defaulters, 81, 126, 223.
De la Rue, celestial photography, 112, 144; his observatory, 154, 188; eclipse 1860, 156.
Delaunay assists in preparing address, 158,; lunar acceleration, 159.
Dembowski, medal, 198.
De Morgan, A., on foundation of Society, 21; on calculations, 26; work as Secretary, 79; on prospects of astronomy, 81; his character, 86; on circumlocution, 122; resigns seat on Council, 141.
De Morgan, William, 30.
Denison, see Becket.
Denning, Observing Society, 135; lists of radiant points, 199.
Devonshire, Duke of, Royal Commission, 174, 207.
Dinner, 2, 21.
Disraeli, his schooldays, 25.
Dissensions in Council, 176 sqq.; in Society, 191.
Donati's Comet, 113.
Downing, Superintendent of Nautical Almanac, 217, 218.
Draper, H., photographs nebula in Orion, 213.

Dublin Observatory, 72.
Dun Echt Observatory, 194; circulars, 220.
Dunkin on motion of solar system, 150.
Durchmusterung, 116, 145; proposed southern, 146; photographic, 213.

Earth, figure, 69; density, 69, 87, 117, 118.
East Sheen, Pearson's School, 24.
Eclipse, total solar, 1842, 108; 1851, 112, 119; 1868, 194; 1869, Gould on, 170; 1870, 164, 169, 170; 1871, 171; 1875, 190; 1878, 199; 1887, 215; 1889, 215; 1893, 216; reports, how published, 216, 241.
Eclipse Committee, 215; joint, 216.
Eclipse Volume, 171.
Ecliptic, position of, Airy, 70.
Edinburgh Observatory, vacancy at, 67.
Encke, medal for orbit of comet, 43; his Jahrbuch, 58; medal for it, 60; orbits of binaries, 75.
Endowment of research, 173, 174, 208.
Ephemerides of satellites and for physical observations of planets, 210.
Everest, testimonial, 99.
Evershed, expeditions to Kashmir, 222.

Farrar bequest, 246.
Fellows, number of, 80, 83, 211, 245; non-resident, 80, 211; foreigners as, 232; women, 233.
Fifty years' celebration, 167.
Fisher, G., pendulum observations in the Arctic, 33.
Flamsteed, copy of correspondence presented by Baily, 65; Baily's book on, 78.
Foster, pendulum observations, 68.
Foundation of Society, 1 sqq., 48.
Founders, 1 sqq., 18.
Fowler, centenary address, 237.
Franklin, John, early member, 33.
Franklin-Adams, charts, 243.
Franz, lunar nomenclature, 223.
Freemason's Tavern, dinner at, 2, 21.

Gasparis, discovery of minor planets, 115.
Gautier, magnetism and sun-spots, 117.
Geodetic work in England, 223; base at Moray Firth, 224; arc in Central Africa, 224.
Geophysics, meetings on, 227; papers on, 228.
George III, death of, 1, 7; South's anecdote about, 55.
Gesellschaft, Astronomische, foundation of, 151.
Gill observes Mars at Ascension, 194; bequest, 195 n., 246; stellar photography, 213; reform of Nautical Almanac Office, 218; methods of determining places of planets, 221; arc in Central Africa, 224.
Glaisher, James, balloon ascents, 145.
Glasgow Observatory, 67.

Glass for object glasses, 16 ; repeal of duty on, 108.

Gorton starts *Astronomical Register*, 134.

Gould, photographs of clusters, 213.

Grant, *History of Physical Astronomy*, 121, 133 ; editor, 133.

Green, drawings of Mars, 197.

Greenwich Observatory, Board of Visitors, 65 ; Airy and, 71.

Gregory, O., founder, 2, 3.

Greig, Admiral, presents altazimuth, 83.

Groombridge, founder, 2 ; catalogue, 65.

Grove-Hills bequest, 246 *n*.

Guinand, glass for objectives, 16, 108.

Hale, address by, 238.

Hall, Maxwell, solar parallax, 197.

Halley's observations, MS. copy of, 64.

Hansen, medal for lunar and planetary theories, 91 ; testimonial, 99.

Hartnup, Assistant Secretary, 63 ; photos of moon, 113.

Harton Colliery, pendulum observations, 117.

Hartwell House, 34 ; Observatory, 76 ; advowson, 81, 124, 200.

Hencke, discovery of Astræa, 92 ; testimonial, 99.

Henderson, computes occultations, 58 *n*.; Astronomer Royal for Scotland, 67 ; on refraction, 70 ; lunar parallax, 70 ; annual parallax, 70, 89 ; justice to, 90.

Henry Brothers, stellar photographs, 213.

Herschel, Alexander, registration of transits, 173 ; meteors, 199.

Herschel, Caroline, Honorary Member, 81 ; her telescope presented, 83.

Herschel, John, diary, 1 ; one of the founders, 2, 3 ; state of science in 1820, 16 ; South's telescope, 52 *n*.; on *Nautical Almanac*, 59 ; observes double stars, 74 ; nebulæ, 74 ; orbits of binaries, 75 ; expedition to the Cape, 75, 106 ; first to use glass negatives, 76; his Cape work, 87, 106 ; on Hyperion, 98 ; testimonial, 99 ; reference Catalogue of Double Stars, 202.

Herschel, William, first President, 11, 25 ; collected papers, 242 ; his MSS. presented, 244.

Higgs, atlas of solar spectum, 243.

Hind, testimonial, 99 ; discovers minor planets, 115 ; edits *Nautical Almanac*, 216.

Honorary Members, 81, 192, 234.

Horrocks, Memorial to, 185 ; library fund, 186.

Howlett, drawings of sun, 153, 192.

Huggins, his observatory, 115, 153 ; spectroscopic work, 152 ; held office for many years, 245.

Hussey, his observatory, 66.

Hyperion, discovery of, 98.

Indian Hill Observatory proposed, 138.

Instrument Committe e, 45 ; instruments lent to Fellows, 76 ; not traced, 188, 190.

Ivory, afraid of enemies, 78 *n*.

Jackson, editor of *Astronomical Register*, 135.

Jackson-Gwilt medal and gift, 236.

Jacob, Indian Hill Observatory, 139.

James, density of earth, 117.

Jansen bequest, 192.

Johnson, catalogue of southern stars, 67 ; Radcliffe Observer, 72.

Joule, mechanical equivalent of heat, 111.

Juno, solar parallax from observations of, 194.

Kapteyn, Cape D. M., 214.

Kashmir, Evershed's expeditions to, 222.

Kelly, founder, 2, 21.

Kent, Duke of, death, 1.

Kew Observatory, 155.

Kirchhoff, solar spectrum, 114.

Knobel, vindicates Smyth's character, 206 ; list of star catalogues, 211.

Lacaille's arc remeasured, 106.

Lambert bequest, 192 *n*.

Lassell, discovers Hyperion, 98 ; medal, 99 ; telescopes, 107, 114.

Latitude, variation of, 225.

Lee, Treasurer, 34 ; presents a circle, 46 ; observatory, 77 ; advowsons, 81, 83, 124, 200 ; benefactor, 81 ; elected President in opposition to Airy, 141 ; life, 142.

Lee fund, 81.

Leonids, 160 *sqq.*

Le Verrier, Neptune, 93 ; testimonial, 99 ; orbit of Leonids, 162 ; planetary tables, 162, 191 ; intra-mercurial planet, 194.

Library, 64, 243 ; catalogue, 64, 244 ; of Spitalfields Mathematical Society, 103.

Lindsay, Lord, 170 ; Mauritius expedition, 194.

Lockyer, spectroscopic work, 154, 169 ; proposed for medal, 172, 198 ; retires from Council, 176 ; his observatory, 188 ; appointed to College of Science, 207.

Longitude, Board of, 15, 56 ; abolished, 58.

Louis XVIII at Hartwell, 34.

Lubbock, planetary theory, 68 ; testimonial, 99.

Lucknow Observatory, 118.

Lunar nomenclature, 222 ; tables of Burckhardt and Hansen compared, 219 ; Theory, 197, 199.

McClean bequest, 246.

Maclear, at Biggleswade, 76 ; at the Cape, 76, 106.

Main, his work, 136 ; address in 1864, 147.

Manners, President, 163.
Manuscripts, 65, 83, 109, 243 n., 244.
Markree Observatory, 189.
Mars, solar parallax from, 194; drawings of, 197.
Maskelyne, edits *Nautical Almanac*, 56.
Mathematical Society at Spitalfields, 99–102, 104.
Maule, arbitrator, 54.
Maxwell, Clerk, 111.
Mayer, Tobias, his catalogue, 77.
Medal, 42; first presented, 43; none for discovery of Neptune, 97; method of bestowing, 125; giving it to two, 172; proposals to abolish, 235.
Medical and Chirurgical Society, rooms at, 35, 60.
Meetings, record of first year's, 11–13; first seven years, 39; reports of, 134, 238; hour of, 229 *sqq.*; special, 238.
Memoirs, regulations for, 36; early volumes, 37; in the thirties, 79; in the fifties, 126; how distributed, 241; appendices to, 242.
Mercury, transit of in 1878, 197.
Meteors, Leonids, 160; connected with comets, 161; observations of, 199; stationary radiants, 199.
Middleton, founder of Mathematical Society, 83, 99; portrait of, 83.
Miller, two of that name, 160.
Millington, Secretary, 42.
Minor planets, 68, 99, 115, 120, 137.
Mitchel, O. M., on South's telescope, 54; Associate, 105.
Mitchell, Maria, discovers a comet, 105.
Monthly Notices, started, 38, 41; in the thirties, 79; enlarged, 84; in the fifties, 127; in 4to, 133, 239; growth of, 188, 240; geophysical supplement, 228; how edited, 239; takes place of *Memoirs*, 239; old volumes scarce, 80, 241.
Moon, tables of, 148, 219.
Moon-culminating stars, 73.
Moore, Daniel, founder, 2, 3, 20.
Müller, Max, on Colebrooke, 18.
Mural Circle, favourite instrument, 72.

Nasmyth, solar surface, 153.
Nautical Almanac, 55; reform of, 56 *sqq.*; committee on, 61; changes proposed in 1890, 216; Society not consulted in 1897, 217; Part I. of, 218; changes from 1914, 218; suggested change in office, 218.
Nebulæ, J. Herschel's observations, 74; W. Herschel's catalogues revised, 243.
Neison, on lunar theory, 197, 199.
Neptune, discovery of, 92–97; attitude of Society to, 97.
Newall, H. F., special address, 237.
Newall, R. S., 25-inch telescope, 189.
Newcomb, lunar theory, 199; Council resolution on his retiring, 219 n.; commencement of astronomical day, 225.
Newton, H. A., on November meteors, 161.

Observatories in southern hemisphere, 66, 221; Indian Hill Observatory proposed, 138; list of, in the sixties, 131; in the seventies, 189; solar observatory in Australia, 222.
Observatory, The, monthly magazine, 198, 238.
Observing Astronomical Society, 135.
Oxford, Radcliffe Observatory, observing at, 72.

Parallax, lunar, Henderson, 70.
Parallax, solar, Encke's value too small, 149; Stone, 163; Gill, from Juno, 194; from Mars, 195; from transit of Venus, 196; Maxwell Hall, from Mars, 197.
Parallax, stellar, Henderson and Bessel, 70, 89.
Paramatta Observatory, 66.
Parry, Edward, proposes additions to *Nautical Almanac*, 61.
Parsons bequest, 246 n.
Patron of Society, 46, 50, 81.
Pearson, founder, 1–3, 23; buys object-glass, 16; earliest idea of founding Society, 21; school at East Sheen, 24; star catalogue, 74; presents stock of his Practical Astronomy, 83.
Peirce, Benjamin, Associate, 105.
Pendulum observations, Foster, 68; Baily, 69; Airy, 117.
Philosophical Magazine, reports of meetings in, 39.
Photographic Committee, 214.
Photography, J. Herschel's glass negatives, 76; daguerreotype of eclipse of 1851, 112; photos of moon, 113; of Donati's comet, 113; De la Rue's work, 144, 156; in Transit of Venus, 181; advance of stellar photography, 213; photos for sale, 215.
Photoheliograph, at Kew, 113, 155; at Greenwich, 156.
Pond, editor of *Nautical Almanac*, 60; work at Greenwich, 71, 73.
Portraits, 83, 244.
Printing, cost of, 40.
Pritchard, President, 159, 162.
Prize questions, 43.
Proctor, 167; on transits of Venus, 168, 180–184; proposed for medal, 176; disavowed by Council, 184.
Prominences, nature of, 164.
Ranyard, 169; eclipse volume, 172; proposes to abolish medals, 235.
Refraction, Atkinson on, 70.
Register, The Astronomical, 134, 238.
Relativity, discussion on, 238.
Reports, Annual, of Council, 240; of meetings, 134, 238.
Robinson, T. R., work at Armagh, 72, 73.
Rosse, third Earl of, on South's telescope, 54; his telescopes, 77, 107, 113; fourth Earl, on moon's heat, 189.
Rümker, C., silver medal, 43; gold medal for star catalogue, 123.
Rutherfurd, photography, 113, 197.

Sabine, magnetism and sun-spots, 117.
Sadler-Smyth, scandal, 201 *sqq.*
Saunder, lunar nomenclature, 223.
Savary, orbits of binary stars, 75.
Scale, standard, 69, 107.
Schönfeld, Southern *D.M.*, 146.
Schumacher, tables, 57, 58; precarious position, 1848–50, 107.
Schwabe, periodicity of sun-spots, 108, 116; drawings presented, 193.
Science, state of, about 1800, 16; about 1850, 110.
Secretaries, in the twenties, 45.
Selwyn, photos of sun presented, 192.
Sheepshanks, R., on foundation of Society, 48; pamphlet against Babbage, 52 *n.*; on South's telescope, 53; Groombridge Catalogue, 65; Cape Mural Circle, 72; edits *Monthly Notices*, 80 *n.*, 84; death, 118.
Sheepshanks, Miss, bequests to Cambridge and to Society, 192.
Shooting stars, see Meteors.
Signatures, book of, 46.
Silvered-glass mirrors, 114.
Slavinski, founder, 2, 32.
Smith, H. J. S., on endowment, 210.
Smyth, Piazzi, eclipse, 1851, 119; expedition to Teneriffe, 120.
Smyth, W. H., Observatory, 76; medal, 92, 202; his work, 142; attack on him by Sadler, 201; vindication by Knobel, 206.
Sniadecky, 33.
Societies, new scientific, 17, 18.
Solar Observatory, 175; in Australia, 222.
Solar Physics, Committee on, 207.
Solar Spectrum, Kirchhoff and Bunsen, 114; Rutherfurd, 198; Higgs, 243.
Solar System, motion of, 149.
Somerset, Duke of, elected President, 7; declines, 7–10.
Somerset House, rooms at, 64; longitude and latitude of, 82.
Somerville, Mrs., Hon. Member, 81, bust of, 83.
South, founder, 2, 21, President, 51; his telescope and lawsuit, 52; on reform of *Nautical Almanac*, 57, 59, 60.
Spectroscopy, stellar, in 1863, 152.
Spitalfields Mathematical Society, 99.
Standard Scale, 69, 107.
Star Constants, 15.
Stokes, C., founder, 2.
Stokes, G. G., change of refrangibility of light, 112; astronomy in 1869, 165.
Stone, advowson of, 83, 124.
Stone, E. J., solar parallax, 163, 196.
Strange, career, 173; on endowment of research, 174, 175; opposition to Council, 177.
Stratford, silver medal, 43; Secretary, 44, 63; Superintendent of *Nautical Almanac*, 62.
Stratosphere, 145.
Struve, O., on discovery of Neptune, 96; medal, 99.

Struve, W., member of committee for reform of *Nautical Almanac*, 61.
Sun-spots, period, 108, 116–117; drawings of, 109; MS. observations belonging to Society, 193.
Surface of sun, Nasmyth's observations, 153.
Survey of heavens proposed, 6, 49.
Survey, Ordnance, 117, 223.
Sussex, Duke of, helps to get rooms at Somerset House, 64.

Tables, requisite for use with *Nautical Almanac*, 65.
Talmage on Sadler-Smyth scandal, 202 *sqq.*
Taylor, Henry, reduces Groombridge's observations, 65.
Taylor, Richard, printer, 40.
Telegrams, astronomical, 220.
Tennant, on Lucknow Observatory, 119; proposes changes in *Nautical Almanac*, 216.
Testimonials awarded in 1848, 98.
Thomson, William, work in thermodynamics, 111.
Transit Ephemerides, 62.
Transit of Venus, see Venus.
Troughton, South's telescope, 52; Groombridge's transit circle, 72.
Tupman, results of Transit of Venus, 196.
Turnor presents MSS., 83.

Union astronomique internationale, 228.
Uranus, motion of, 66 *n.*, 92.
Usherwood, photo of Donati's comet, 113.

Venus, Transit of, preparations for observing, 168, 178–184; methods of observing, 179; observations in 1874, 185; results, 196.
Visitors admitted, 234.
Voting by proxy or by post, 231.
Vulcan, alleged intra-mercurial planet, 193.

Walker, S. C., Associate, 105.
War, subscriptions during, 233.
Waterston, on kinetic theory of gases, 27; other papers, 27–28, 49.
Weale, publisher, 40.
Weisse, testimonial, 99.
Wellington, inauguration of South's telescope, 53.
Wesley, W. H., Assistant Secretary, 187; map of moon, 223.
Whitbread, efficient Treasurer, 125, 137.
Wilcox, Director of Lucknow Observatory, 118.
Williams, John, Assistant Secretary, 84, 186.
Wolf, R., period of sun-spots, 117.
Wollaston, W. H., presents telescope, 45.
Women as Fellows, 233.
Wrottesley, star catalogue, 73.

Yorkshire Philosophical Society, 23.
Young, Thomas, resists reform of *Nautical Almanac*, 56–60.

PRINTED IN GREAT BRITAIN BY NEILL AND CO., LTD., EDINBURGH.